I Am
Who I am

KENNETH WALDEN

ISBN 978-1-0980-3176-3 (paperback)
ISBN 978-1-0980-3177-0 (digital)

Christian Faith Publishing, Inc.
832 Park Avenue
Meadville, PA 16335
www.christianfaithpublishing.com

Printed in the United States of America

INTRODUCTION

The most important person in my life was and still is my grandmother. She raised me to be all the good I have in me and cautioned me to fight the demons that would appear all throughout my life. In this biography, you will see me refer to her. This is my way of showing what can happen to you when you believe that there is an answer to all the painful problems that you have endured. I have endured many and have found the answer, which I am sharing here. I hope you read this. You may find something for yourself.

CHAPTER 1

The Beginning Years

My grandparents were born around 1875 in North Carolina, both on a slave plantation. They migrated to Boston, Massachusetts and brought a four family house in 61/ 63 Ottawa Street, Roxbury, Massachusetts, in the 1920s. They raised three sons and two daughters. Plus, when my grandfather died, she raised me and my uncle's daughter Joan. This story is a tribute to her love, patience, and understanding.

I was born on April 3, 1942 at Boston City Hospital. My grandmother raised me, teaching me all I needed to know to get through all the things that I have endured. She believed that praying to God will help you find your way. However, my view has always been that God helps those who help themselves. Of course, I never got up enough nerve to tell her what I thought.

I want to first introduce you to my immediate family. My father was Dr. Robert Edison Walden. My mother was Gertrude Bradshaw Walden. My grandparents were Charles E. Walden, Mary E. James Walden. There was my uncle Charles E. Walden Jr., his wife, Fannie Walden; my uncle Raymond E. Walden, his wife, Bertina Walden; my aunt

Adelaide E. Walden; and my aunt Eleanor E. Walden—all with the middle letter E, which I thought was to identify us, maybe as a result of my grandparents' slavery days.

It is important to note that most families that lived in my neighborhood came up from the South and were born around the time of Reconstruction, facing impossible odds coming North to find jobs, working and saving every penny they could, raising children, and putting up with all the prejudicial treatment, and then being able to buy their own homes.

No one in my neighborhood locked their doors because everybody knows everybody. My grandmother was called Miss Mary or Mary E. by everyone. I guess it was a sign of respect because many in the neighborhood called upon her to provide nursing care to all who were sick or were slightly injured because going to the hospital meant all of us blacks who lived in the area had to stand in the back of the line in order to get care at Boston City Hospital. You may have been lucky to come to Boston, but your treatment was still very much the same, only not in the open. You were fine as long as you stayed in your place.

She once told me that when they cooked for the big house, they left us a lot of sauce on the vegetables. When the meal was over, the household help was given the leftovers. Little did the owners know that the leftovers had all the vitamins, especially in the juice. So when winter came, the white folks would become sick while us slaves ran around with little clothes on and remained healthy.

My street was lined with four family houses all looking the same, so much so that if you didn't know exactly where you lived, you could go into the neighbor's house. We had a pig nut tree in the front yard of our house; it's probably still there. We lived on the second floor with a big kitchen, a big dining room, a big living room, a side room that had a piano, porch out front, and upstairs with four bedrooms. Our home was heated by cold furnaces and a black belly coal-burning stove in the kitchen, with a gas water copper boiler, which was necessary to heat water for the whole house. We had two big scrub sinks, and the bathroom had the old-fashioned flushing toilet.

My uncle Charles had seven children named Joan, DeeDee, Norma, Richard, Robert, Freddy, and Raymond (called Ray-Ray, who was born much later). They lived on Hazelwood Street just a couple of blocks from where we lived. My cousin Joan lived at my house. She was about ten years older than me. I remember that she would always find an excuse not to help out with the dishes after we ate.

My uncle Charles always cut my hair. He used the old-fashioned clippers. No style, just cut close. The good thing was, before I left, he always gave me a little change. I never knew why, but I loved getting it. My uncle Raymond and his wife, Bettina, lived next door in the same type of house that I lived. Rent was $42 for upstairs apartments and $28 for the downstairs one-bedroom apartment, which was where Mr. and Mrs. Douglas stayed with their son Sonny, who was my cousin Joan's age.

My grandfather worked at Dudley Street Station as a switchman operator on the elevated train track booth, where the train from Dudley Street went to Egleston Station. It was the MBTA at that time. My aunt Eleanor worked in Washington, DC, at the archives building (Appendix 1 talks about her). She was the only black woman to be noticed for her work in cataloging department. My aunt Adelaide work in Baltimore at the munitions factory during the war. After working as an LPN at Jewish Memorial Hospital on Townsend Street Roxbury Massachusetts, my uncle Raymond worked at Wentworth institute for about thirty years. My aunt Bertina worked at Filene's for about twenty-five years. My father was attending medical school at Lincoln Medical College during those war years. My mother was there with him also; however, I'm not sure what she was going to school for.

My earliest memory was being bathed in the kitchen tub, as it was called then. It was made out of the same stone that we children used to draw lines for hopscotch. I do remember one important thing: when I became old enough to wash myself, my grandma would always ask if I had washed behind my ears. I would always say yes, and then she would call me and put her finger behind my ear and then smell it and say, "Boy, you didn't wash them right. Go back and do it again." I couldn't understand how she knew until I had my own children. I would say the same thing that my grandma did. The fact of the matter was, a person always sweats behind their ears and leaves an odor.

The next thing that comes to mind is my grandfather giving me a whipping for not coming straight home from the store with the newspaper. I was being educated that when you go to the store, you came right home before you do anything else—a lesson that I learned right away and never repeated it.

I believe I was about five years old. My grandfather had died around that time, and my grandmother took custody of me from that point on. During that time, my grandmother was showing me how to wash clothes at the kitchen sink using a scrub board and brown Kirkman soap. Bluing was used in the water to make sheets white. She also showed me how to iron clothes and sew buttons on shirts and darn socks. The following year, I was washing the kitchen floor with brown Kirkman soap and a scrub brush, getting down on my knees because my grandmother said, "You must meet your work." She did not believe in using a mop. I remember that kitchen floor being white and red squares like a checkerboard. Also she would cut the sleeves of sweaters and old shirts to have for the winter when the house was cold at night for us to wear when we went to bed at night because the house would become cold after the furnace went down.

My grandmother called me Chinky, a name that my sister and brothers still call me because my eyes always slant when I am upset or smile. She used to tell me that I would grow up with a special gift, so I would have to be very patient and learn all the things I could and know how and when to use them later. I was much too young to understand at that time. I did pay close attention, watching and learning everything. So when I became an adult, it became very important later in life.

She once told me a story of her relative, Rubin James, born in Delaware who, as a seaman in 1804, fought for this country in the battle of the Barbary Coast. I didn't understand the black history of my people until much later, but now I am proud to know and talk about it (Appendix 2).

I went to Julia Ward Howe Grade School on Dale Street in Roxbury, which was a few blocks from my house. I had a dog called Pal. He followed me almost everywhere. He would do his own shopping at Percy meat market on the corner of Laurel and Humboldt Avenue. He always went to school with me and waited for me to go home.

By the time I was seven, I would catch the trolley car for Dudley Street where I went to pay all my grandma's utility bills (Edison gas and electric bill and Bell telephone bill), most of the time riding on the back of the street car to save the five cents. All of us kids did that because five cents was like a dollar today.

We had a garden in the backyard. Every April, soon after my birthday, it was time to tend the garden. This meant taking a pitchfork and turning the soil. We collected cans and cut out both ends so when they were in the ground, the cut worms couldn't eat the roots. We took our beanpoles out of the woodshed and put them in the ground to plant the string beans in the can so the vine would grow up the pole. We planted tomatoes and string beans the same way. We had a grapevine, an apple tree, and robarb vines. I enjoyed watching our food grow. I would take the hose and water the garden every day. I used to pretend that I was a farmer. I still have a warm feeling thinking about those days. It brought me even closer to my grandma.

There was a split school system at that time where you came home for lunch, and you went back to school. I was wearing short pants at that time when I started school and got into fights. However, my grandmother used to tell me that I better win the fight out there because if I lost, I would definitely lose fights at home. I even started to play the violin, which I hated, but my father went there and played, so I was obliged to go the same route. All my family went to the same schools and had the same teachers all the way through junior high; all participated in some form of music. Because of this, I had no choice. At school, we even had to learn square dancing. Can you see black children dancing square dances, reminiscent to what was happening during slavery days? I was constantly getting in trouble at school and got my hands whipped with the old famous ruler. I was very hard to discipline. Apparently, it was my nature or the devil in me.

When I came home in the afternoon for lunch, my grandmother was always in the kitchen listening to the baseball game. Her favorite team was the Boston Braves. Or she was ironing or doing some kind of housework She was always busy doing something. She would say an idle mind is the devil's workshop. I was never idle, but the devil was always close by.

In 1947, there was a serious snow blizzard. I got so sick that my grandmother decided to take me to Boston City Hospital. The only way she could get there is by putting on a sled, and she and my aunt Berttina pulled me down to Boston City Hospital all through the storm. It was about five miles from our house. When we arrived, the doctor wouldn't see me right away, so my grandmother demanded that the doctor see me. He examined me and pronounced me dead. My grandmother told him, "He is not dead. God hasn't called him yet. Otherwise, He would tell me. You keep working on him." I was told after I recovered they had to quarantine me on the ward because I broke out with pneumonia, chickenpox, and measles. My grandmother told them that it was God's curse for the way they treated me (appendix 4 will tell you about that winter storm).

All these things my grandmother and my aunt told me after I had come home. She would say quite often I had much to do, and God would not allow me to die. Things like this was a sign. I have to admit, all the years I was around my grandmother, I was amazed that she was never wrong.

On one occasion, my grandmother took me to my first and only baseball game. I believe the Boston Braves were playing the Red Sox, and she was definitely a Boston Braves fan. We were supposed to sit up in the bleachers where all black people had to sit. However, my grandmother said no, we weren't. She had bought her tickets, so therefore we could sit down front. She argued with the ticket agent, and don't you know it, we were able to sit down with all the white people. This was always the way my grandmother demanded respect. She educated me that if you don't respect yourself, you will never get respect from others. I don't recall who won the game, but I know my grandmother knew all about baseball. Outdoor playtime was usually one hour after school on weekdays and between two and three hours on Friday and Saturday—if you didn't have chores to do, which I often had. I mostly played basketball from the first time I can remember, beginning with a fruit basket on the telephone pole. Even going to the store, I would have my basketball with me and my dog Pal.

I do recall our insurance man coming and collecting our monthly life insurance. A substitute man came and didn't remove his hat when he crossed the threshold. My grandma knocked his hat off and said in a strong voice that no man should keep his hat on when he entered her house. I believe I was so amazed that I just stood there with my mouth open. After he apologized and left, I asked her about it, and she said to never be disrespected. When the regular man returned, he apologized. I recall our evening walks around the block on Friday and Saturday, and we'd say hello to neighbors who were sitting on their porches, stopping for a moment to chitchat; and quite often, my grandmother would tell me to climb over the front fence and take a piece of the neighbor's plant. This was a normal, neighborly thing back then. You couldn't do that today, not even taking a walk—period.

Grandmother pointed out the importance of doing your work or anything to the best of your ability, to never blame or point a finger at anyone if you got yourself into trouble because you must look at the other three fingers—it was your decision. Most importantly, when you take on a task, see it through, no matter if you are appreciated or not because God takes note.

I attended St Mark's Church on Townsend Street, Roxbury, every Sunday morning for Bible study at 9:00 a.m., came home, and went back to church with my grandmother at 11:00 a.m. for church services. When I got good enough, my grandmother had me play the church hymns in church. This always brought her so much joy.

After church, we went around to the sick members' houses. She was a member of the Rosary Club. At their home, I would play the church hymn sit in the corner, and I had cookies and milk and stayed quiet until my grandmother was ready to leave.

The golden rule was that you never spoke until you were spoken to. I broke that rule many times. For some reason, I always had something to say. I've got so many slaps with the back of her hand on my mouth; I thought that I was a punching bag. On some rare occasions, she was glad that I did speak. One thing that she said was, "Never open up your mouth and speak unless you have something important to say. Only fools run their mouth. You learn more by listening instead of opening up your mouth." She constantly told the people that we met that if they needed anything done to just call her, and her grandson would come and do it gladly. I thought that I worked for "rent a child" services. One thing I finally learned, it was my grandmother's way of keeping me busy so that I wouldn't have the devil playing with my mind. However, that didn't work. Somehow the devil always got his due. I felt I couldn't help myself. Or at least that's what I told myself. She would tell me that God keeps a book on all of us, so when our number was called, we'd have to answer for what he has recorded. So it was important as I became an adult to keep the ink on the right side of the book wet; in that way, I may stand a chance not being sent downstairs.

There was one thing about church that had always disturbed me. During this time, we had a telephone that was a party line. This meant that you shared your line with others. I still remember our numbers. Garrison 7-3231. On the weekend, while my grandma would be taking her nap, I would listen in on our neighbors' conversations. Many times I would hear them talk about another churchgoing member's personal problems, quite often saying that Sister Someone's daughter was going South to have her stomach sickness taken care of. Then on Sunday they would greet the same member as a close friend like they weren't talking about them while praying to God for forgiveness. The funniest thing was, the very next week, they would go right back to doing the same thing. I've always wanted to talk to my grandma about this but knew it would cause me more trouble than good. I wasn't that crazy. This was one of many reasons church didn't appeal to me. But I kept my thoughts to myself for the most part.

During the winter when there was a lot of snowstorms, all of us kids in our neighborhood had sleds that we took to Monroe Park behind the Lewis Junior High School, which was the perfect place to go. I always went to Hazelwood Street where my uncle Charles lived. There was a steep hill there with a blind side because of the wall that ran around my uncle's house. We always had someone at the bottom to tell us when it was safe to come down—except for one day. I was coming down, and my friend was too busy talking to signal. When I was on my way down, a green telephone-repair truck was coming down Monroe Street. I went right under and outside the other side without being hit. I don't know why I wasn't killed. However, being me or because I was too young to care, I went back up the hill and waited for my turn to go again.

We also made a little money shoveling the neighbors' sidewalks. I always had to shovel ours and our next-door neighbor's for free and put used coal asses on the sidewalk after. I could understand doing ours, but theirs? I hated it mostly because they had several men in their family, and it was free. I never had the guts to ask why. I wanted to go out and play afterward. Smart move, right?

My aunt Adelaide and uncle Raymond was causing me so much trouble, saying my mother was a street whore and saying I wasn't a Walden. I didn't first understand. This statement has always been a thorn in my side, even today. It raises its ugly head every time my stepmother says anything related to my past in a way that doesn't include my problems, only focusing on my not accepting her or my father's love, acting like my feelings didn't count or didn't happen. Like many people who only see the part that they will accept, I have learned to endure it, but still it's painful. On a few occasions, I tried to bring it up but got nowhere. That was the way it is in my family's house. I have come to realize that the whole truth doesn't matter if it changes your understanding.

My aunt even threw a meat fork at me one evening when I was sitting at the kitchen table eating. It struck my arm. The marks are still there. My grandmother happened to not be home. Soon I heard it was because my mother got pregnant in college before she got married. It was supposed it was because she had gotten raped and maybe it's the reason my eyes were slanted.

My grandma did all she could to protect me from that kind of talk. Even when my parents came home to visit and I told them what was being said, I don't recall that anything was ever done about it. Not even my mother, as I recall, did or say anything when they would come home to visit. The strange thing about it was when my father remarried. This subject came up again, so my stepmother knew about this problem. I am not trying to insult or harm any relationship, but this is the truth and must be told to keep this painful fact straight.

My aunt and uncle complained that they were putting my father through college. I guess it meant they had a right to speak their minds. It didn't help that when they were home, they didn't spend much time with me. They always brought a friend home. Sometimes it was Roscoe Lee Browne who went to the same school, Lincoln University. My parents spent their time enjoying their old friends while I had to go to my room and read a book. After a while, I stopped wanting them to visit and refused to come downstairs and see them off. Looking back, I believe this was the beginning of my wanting to be independent and not become close to anyone other than my grandmother. My grandma would try and comfort me the best way she could. But the bottom line is, I never got over it, even today. But because I have lived and learned more about life, I do have a better understanding. However, the pain is never far away. I was always told that nothing happens without a reason. I am still waiting for the answer.

At this time, I was learning how to stoke the furnaces on both sides in the cellar and bring up coal to back doors of every apartment. I had to be up at five o'clock every morning, something I still do today. I never knew what it was like to get up in a warm house. Because it was always my job to make the house warm for everyone, it was also my job to provide hot water for the house. Every Saturday morning, my chores consisted of cleaning the front and back stairs all the way down to the sidewalk. In the summer, I had to wash the windows inside and outside. I polished all the furniture, especially the dining room for Sunday dinner, and washed and waxed the kitchen floor. There was also my usual task of starting the coal furnaces and starting our coal stove for heating the kitchen. We had the old-fashioned radiators that would hiss when the hot water was rising to the boiling point. I remember that they were always covered over with some type of metal plating. When it hissed too much, there was a knob to turn it off.

Every holiday, it was my job to hang the American flag on the porch. It was something that, as time went by, I couldn't understand, or why my grandma was so patriotic when I was being told so many horrible stories about their mistreatment. This is one area I felt it wasn't right that we pay so much respect to a country that has treated my grandparents so terribly. When I asked, I was told that there are some things that if you don't have the power to change then you must go along even if you disagree. She said it is wiser to keep your mouth closed until you have the power to change things then to call attention to the fact that you are upset. Silence is sometimes your best weapon to be used when you have the strength to make changes. I guess she was right because they had gotten this far.

You had to be on time for dinner; otherwise, you wouldn't eat. Meals weren't saved for you, unless you were working late. My best meal was on Saturday night. Grandma would cook my favorite meal: the old-fashioned Boston baked beans full of molasses and hot dogs or a little piece of ground steak that she would buy at the A&P market on Warren Street where we went every Saturday morning along with Kavinor's bakery on Blue Hill Avenue for their day-old bread. On Saturday afternoon, my grandmother would sometimes prepare to bake homemade bread and rolls for Sunday dinner. My cousin Joan and I would be tasked with setting the table for Sunday dinner, including pulling out both ends of the mahogany table and putting in the extending piece. We took the silverware from the mahogany chest of drawers that was opposite the dining room table. I cleaned the table and chest every weekend with Old English oil. We put the white lace tablecloth with a clear plastic over the table and placed the gold-trim plate setting that included a dinner soup plate, bread plate, silver knives, forks, and two spoons with the letter W engraved. There were tall china drinking glasses, engraved napkins, a large meat platter and veggies bowls, plus a gravy bowl with a silver spoon, and finally silver candleholders. I polished them every Saturday. As I got older, I wondered where my grandma got these expensive items. But I never asked. I came to realize that all black people at that time had these things as a show of freedom, a symbol of the big white man's house. This is a perfect example of the independence that we black people had then but have lost today.

My grandma never allowed any kitchen cooking ware into the dining room. After dinner, Joan and I cleared the table, washed, dried, and put away everything. Plus, we cleared the dining-room table and finally swept the kitchen and the dining room floor. That was my grandmother's rule. She cooked, and you washed and cleaned up everything afterward.

I always sat next to my grandmother at Sunday dinnertime. The church minister would oftentimes come right at Sunday dinnertime. I would get very upset about it. I finally had the need to confront him and asked him, Wasn't he married? Why did he always come to my grandmother's house just at dinnertime? Everybody at the table looked at me with amazement, and my grandmother gave me the back of her hand. The preacher left and never returned. Later she told me that I was right, but I spoke out of turn. Plus, he always sat in my grandfather's chair at the other end of the table where nobody sat. There was always a place setting there—a symbol of respect. Even at my family's home today, we do that for our father.

I always had to say what I thought even if it got me into trouble. Plus, I thought that he was taking advantage of my grandmother. Plus, he would always take the last piece of meat, so you know that disturbed me. That's the way I was and still am today. Not saying I was right but admitting that's me.

I remember having a hundred and fifty people that I sold Sunday newspapers to, along with cigarettes that they wanted. I recall the cigarettes at that time were Viceroy, Philip Morris, Kool, and Camel. They were $0.35 a pack. Along with the Sunday newspaper, that was $0.25 cents: Boston Globe, Herald, and Boston Post. I was up on Sunday mornings before 4:00 a.m. because I had to be finished by eight to get ready to go to Sunday school, something that I could never understand because I spent so much time reading the Bible to my grandma. But I wasn't foolish enough to question.

Most of my customers were church members. I was always paid with the money being left under the mat at the front door of the apartments or houses in a white envelope. At Christmastime, there was always some extra money in the envelope. My cousin Freddy helped me sell Sunday papers.

I remember one Christmas when my grandmother bought me a Shelby bike. I was very happy to get it. I was the only one on my block that had a bike. One of the reasons I thought so was because on Easter, most of the time, I only got my shoes resoled and taps put on the end. I had problems with my feet, so I had to wear special shoes from Edward A. Chase store. My feet always grew faster than the rest of me.

I used to shine shoes on the corner of Bower Street and Humboldt Avenue in front of a Chinese drugstore. Also, in front of Slade's on Tremont Street in Roxbury was where Malcolm Little, better known as Malcolm X, used to go, where all black men wore what is called zoot suit, with their watch change dangling out of their pocket, and they had felt hats. It was the style back then that men used a woman's stocking, tying the end and putting it on your head to make your hair show waves. Of course, you had to use a lot of hair grease such as King Kong or conk your hair with the LY hair cream that would straighten out your hair for a while if your scalp didn't burn up. I think lately they called it a do-rag. This was the style of dress in those days.

When I got much older, around fourteen, I took a young lady out on a date You always had to dress like you're going to church. Walking on the outside, opening up the door, holding her chair, helping her with her coat, and making sure you had her home on time—it sounds like work, but it was the way, and you weren't even guaranteed that you were going to get a quick kiss. Need I tell you that not too long after, I surely changed all that. Of course, not without consent.

My aunt Bertina's brother used to play jazz at Slade's playing a sax. That's where I got my education on the street life. All the money I made during this time I've used to bring home and give it to my grandmother. Giving it to her was safer than any bank I knew. She would say on occasion that I would always find a way to make a penny; also she would always remind me to "watch your pennies, and the dollars will take care of themselves."

In the summer on Saturday morning, my grandmother and I would go over to Northeastern University where there was a railroad and a coal bin. During the summertime, people came and got coal chunks for the stoves. It was free, and people had to take advantage of every chance they could get to save a penny. Everyone had some kind of card that they put in a window that the coal man and iceman could read to determine what the person needed at a particular time. We had an icebox. There was a milk delivery truck. Sometimes I would take their return bottles and cash them at my neighborhood store for five cents. Well, they were just standing there calling me, I thought.

When we got our delivery of coal for the furnaces, my grandmother taught me how to go downstairs and watch what was being done so that I would recognize whether any mistakes were made. It was part of the rule back then that if you wanted something done right, you did it yourself—which also meant you never trusted anybody, especially when it came to your money.

During this time, the circus and rodeo would come to town and store their freight cars where Northeastern University is today. We could get free tickets by following the elephants and horses down Tremont Street to the Boston garden where they would perform by shoveling the elephants' waste. I remember seeing Roy Rogers once with his horse Trigger and his wife, Dale Evens, and Gene Autry. That was like seeing the Beatles or Elvis Presley.

I took my red wagon down to Dudley Street and to bring groceries home from Blair's market for people who were down there shopping. I got a few pennies for doing it, and nobody locked the doors. If you wanted some spending money, you had to learn very quickly how to make it honestly. Respect of your neighbors was the most important thing that made our neighborhood safe. You can't find that today.

The movies cost $0.20; plus, the movies had a double feature and a chapter, mostly Buck Rogers, Tom Mix, or Lash LaRue. Milk was $0.15 cents; plus, you could cash in the bottles for $0.50 cents. And Coke bottles were for $0.50 cents, bread was $0.15, and candy was a penny. Speaking of the movies, every year my grandmother had to see Gone with the Wind. I used to go there with her on Friday night to see it. She would always bring homemade popcorn. I must have seen it for many years. I was able to sit there leaning on her shoulder, halfway going to sleep and memorizing every line in the story. I can still remember it. I guess it was a reminder of her time down South and during the Reconstruction.

There wasn't a day that she didn't use her education in life teaching me, using riddles that were meant to make me think about what she was pointing out. All this was from her experience growing up in the South. I could tell you there's nothing more important than getting firsthand knowledge of your past history from somebody who has lived it. This picture was the closest I came to recognizing that period, and it served me later up to today. My grandma eventually went to the commercial school in Warren Street to learn how to read well and write. I used to walk her there and pick her up three times a week in the summertime. However, let me point out that she never lacked for understanding anything before going to this school. She would say that to survive, you have to keep up with the times because time waits for no man.

My grandma had a suitor (male acquaintance) named Mr. Warren, who I think came from the same place and time as my grandparents. He would call her Mrs. E and come on some Saturday mornings always bringing some flowers. He would come with his wheel cart and call up at our kitchen window, saying, "Woo, woo!" and she would come to the widow and respond. He would always ask if she needed any work done. She would always say, "No, I have my grandson to do anything I need done." We would always tease her about him, and she would respond by saying, "It's none of your business." I can't remember what happened to him, but I do remember that he was a knife sharpener and the neighborhood's fix-it man.

My grandmother always had fun playing cards with me. Her favorite games were "war" and "steal the old man's pack." She was extremely good at playing checkers; that's a game I could never beat her at. She had a favorite chair in the dining room that she sat in, and I sat on the floor by her where, on Saturday night, we would listen to Amos 'n' Andy, Jack Benny with his black servant Rochester, The Shadow Knows, and Inner Sanctum.

She loved to listen to me read to her the Old Testament. She was partial to the Old Testament because she used to tell me that the Old Testament was the Word of God. The New Testament represented those people who talked about what they were told about God through others. So if you want to know the words that God spoke, you read the Old Testament. It took some time, but I finally got up the courage to ask her why we went to St. Mark's Church, where they talked all the time about Jesus and the New Testament, and they barely mentioned God in the Old Testament. When the Bible says worship no other God but him, she said that she was aware; however, as long as she knew the difference, you didn't mind listening to what Jesus was saying. Then she pointed out that in life, you have to always listen to somebody else's point of view and keep your decisions sometimes to yourself if you feel it's going to cause a problem. A message that I always tried but most of the time couldn't follow in life because I had a habit of saying what I thought. Maybe because I was an Aries. Or the devil made me do it

When we finally got a TV, it was a big box Olympic twelve-inch with an antenna where you had to wrap an aluminum foil around it. She would always watch Ed Sullivan, MSquared, and Jackie Gleason. Sometime after around 11:00 p.m., the station would close down with a sigh on the screen. She never let me turn it on until the June Taylor dancers were finished because she didn't want to let me see them kicking up their legs. I used to even turn our dining-room clock ahead, but it never worked. She had an inner clock that she went by, something she taught me because I have never needed to have one to know the approximate time. When I was sent to bed, I always waited because I knew that the TV would always need fixing, and I was the only one who could use the old-fashioned fix-it method of hitting it on the side to make the picture come back; then I would slide in beside her, half-sleeping, and watch more TV. Boy, were those the days.

CHAPTER 2

Adventure

My grandmother had a nephew named Chris, who was in the navy and would come visit every now and then from Maine where he worked at a lighthouse. He would always come in drinking and saying, "First today." My grandmother used to always try and stop him; however, she couldn't do anything about it. I picked it up, and when I would say it while holding a glass of milk, she would pop me in the mouth with her dishcloth. I believe she could pop the draws off a fly. That was a legal weapon in her hands.

One winter she let me go with Chris to Maine for a visit. I remember there was a lot of snow around the lighthouse. In the morning, I got up and heard something at the door.

Me being nosey, I opened it up and was shocked to see the biggest moose you've ever seen. I couldn't move. I even think I wet my pants. It seems like it took an hour for it and me to decide who was moving first. I finally backed off, and I came to my senses and shouted and locked the door. I woke up Chris and told him what had happened. He just sat there and laughed, then said, "Oh yeah, I should have told you that Fester my friend who drops by. He was checking to see

24

if I was home." You know, I was ready to go home. It was an adventure, but I never returned.

During that same period, my grandma took me to Denver Colorado, to visit her sister. We went by train, the coal-driving locomotives that had a combination of freight cars and passenger cars. We had a sleeper car in the back where black people rode, and the black smoke from the coal coming from the engine could be smelled. Grandma and I slept together, me on the window side so I could look out and see the scenes. We had Wonder Bread and fried chicken, plus peanut butter and jelly sandwiches to eat because there was no food service for us. I do remember going around the famous Horseshoe Curve.

When we got there, my aunt gave me a cap gun. I drove everyone crazy shooting it, but no one said much about it because there wasn't anything else for me to do; many people still didn't have a television. I do recall looking out her back door and seeing the mountains with the snow on top. We stayed for about two weeks. When we left, I got a new pair of cowboy boots. This was an exciting time.

At the end of this year, trouble came knocking again. First I took upon myself to settle an old score. I took my cousin's BB gun that my cousin stored in our basement and waited till my grandmother wasn't home and shot up my aunt and uncle. They ran outside and called the police about that time my grandma came home. She came in and took my gun and told the police generally what it was all about. Back in those days, there was no such thing as black children pro-tection—at least we never had it—so the police left but took the gun. I remember my grandma saying to them, "Maybe now you will leave him alone," and then rubbed the top of my head and smiled, something I always loved.

CHAPTER 3

Bad Memories

It wasn't long after that she talked to me about going to see my father and mother and getting to know my mother, who was now living in Coffeyville, Kansas. I really didn't want to go, but my grandmother asked me, so I knew I had to go. I flew out to Kansas City on my own, and my father met me at the airport. We went to a college friend of his in Kansas to spend the night before we were ready to drive to Coffeeville.

While there, I made a big mistake by taking my father's keys off the table while he was in another room talking and went down to start his standard shift car. It jumped the curb on the oil pan and started leaking, so we took a week to have it repaired. For some reason, he didn't beat me; however, we had a serious discussion. I remember that at that time, I was able to get a lot of my pain off my chest. He was a country doctor at that time in Coffeyville, Kansas, getting paid by chicken and vegetables mostly, as I remember.

My mother was there with my newly born sister Roberta, telling me that if it wasn't for me being born, she could have a better life with her husband. I never forgot those words. I thank God I only stayed a week because it was miserable. I came home thinking I would never return again. My grandmother told me when I got back she had hoped that things would work out; however, she felt sad for me.

I remember her talking to my father, and I overheard her saying, "It isn't right for Chinky to be treated this way, and it was a big mistake having him go there." All these things made me change my view of everything and made it much more difficult for me to get along with people in general, especially in school. Things like this affected me even today. The following year, as I remember, my mother and father got divorced.

As much as I tried, with my grandma trying to help, I began to change—not that you could see it, but my grandma could tell. I still went to church but not to early-morning study. I still played my violin and was now taking communion The minister was still taking three collections, always having something else going on. However, it was about to come to light, and the deacons of the church discovered that the minister had been dipping into the pot. I felt this vindicated me of what I had felt and said much earlier. My grandma, who was also a deacon, became very upset. I remember reminding her that the Bible says to forgive the sinner. She and I were both surprised that I remembered and said that. I never quite understood why myself.

CHAPTER 4

Starting to Mature

It was around this time that my aunt Adelaide got married to Herald Wilson, a long-distance truck driver driving a Mack 18-wheeler diesel who worked for E. J. Scannel, located in Somerville, Massachusetts. He soon played a very important part in my life.

Now it was time for me to make a major change in my life. Herald began taking me on trips to Washington, DC. This happened during summer vacation and holidays. I learned everything about trailer trucks and drivers, what the road signals were, and most importantly, what it meant to be a union member. It was around 1950. I saw Jimmy Hoffa at a meeting. These meetings got heated. On a couple of occasions, I had to hide under the dashboard because there were fights. My uncle told me to keep quiet about what I saw, so I did. At the time, I thought it was a great adventure, not understanding how serious it was.

I learned that they would use their lights to talk to one another. They didn't have two-way radios then. They would blink their lights when they passed and went to get back in line. Also trucks going the other way would signal your truck if there were police traps ahead, and finally, what was most interesting was how they would get behind one another, sometimes four or five in a row, and push one another up the Seven Sisters. We always stopped at truckers' rest stops where truck drivers from companies like M & M, St. Johnsbury, and others gathered to talk about their union. I never thought that this experience would be very important for me later in life. I believe the only other truck company was Peterbilt.

At home when he came with his truck all ready for his trip, it would interrupt all the televisions on the street when the motor was running. When it was close to the time for him to leave, he always asked me to go and start it up. It would cause so much trouble for our neighbors' homes, which shook and interfered with their TV. I would rev up the engine so everyone would know it was me. They would complain, but I always thought it was fun because my friends would know I was behind the wheel. Boy, you know that was fun. Little did I know that the devil was watching because it wasn't long after that I started borrowing my neighbors' cars with a General Motors key that I found. During those days, one General Motors key almost fit all General Motors cars.

When he wasn't on the road, Herald came home driving his new Buick Roadmaster 75. This was a high-end car at that time, fully equipped with all the latest features. I would always wash and wax his car as needed using Simoniz wax. This was the best wax at that time. I also did the same to his cab. I learned that truckers took better care of their cabs than their wives because, as they said, "Your cab is your lifeline. Your wife is just your wife." This allowed me to try out my newfound key. He was the only one that knew I had it at that time. I always had to move his car to wash the tires.

I was seen by all the other kids on the street as I pretended that I was driving—showing off, I guess. All these things kept me from focusing on my mental problems. It was like a God-sent relief. My grandma, I believe, noticed that I wasn't as upset as before, so I guess she allowed me to enjoy my newfound adventure riding with Herald, but not me driving. However, here came the devil, taking charge.

I started going out my bedroom window late at night and borrowing neighbors' cars, driving around the block, and then putting it back in the same spot. I got away with it until I got greedy and started to show off in the late afternoon or in the evening. I would take a car up Humboldt Avenue to Franklin Park and turn around and come back, with my left arm out the window like I would see all the older black men do. It was called the gangster lean. What did I know? You have to live to learn, right?

The police would come to my house every time a car was gone, thinking I had it, even when it wasn't me. You know, that didn't make my grandmother happy. So I did slow down, but I couldn't stop completely. Once I even took Herald's car from the trucking company parking lot in Summerville and drove it back to Roxbury; then I took my friends for a ride. Then I was told that my aunt found out and was looking for me. So I drove the car back to Summerville, parked it in the same spot, and left. When I got home, she said that she couldn't prove I had it, but the hood of the car was warm, and her husband wasn't the driver; plus, no one saw anything. I know some of the truck drivers saw me because I waved at some of them when I was there. So I felt they were protecting me like a union truck driver. When Herald got home, he quickly told me he knew but didn't get upset. We would sometimes stop by my aunt Eleanor's house in DC, so I remembered where she lived for later. Eventually, he and my aunt had two children by the name of Herald Jr. and Karen, who, as time went by, stuck to me like glue. They moved to Crawford Street near Walnut Avenue.

CHAPTER 5

Getting Too Big for My Britches

While attending Lewis Junior High, I had gotten into trouble for not wanting to say Columbus discovered America—mainly because he never came here, and the Native Americans were already here. The school sent me home. Big mistake. My grandma stopped what she was doing, put on her everyday coat and hat, and went to Lewis Junior High, moving faster than I had ever seen her move. Once there, she went right into the principal's office, whose name was Mr. Guindon, and said, "My grandson knows his history because I teach him. So if he said Columbus didn't discover America, he must be right. How can anyone discover some place if someone is already there? So I want him back in class." When she came up for air, the principal said, "Yes, Ms. Walden," and that was that.

I was still going to school. I stood about five feet eight inches. I had been playing a lot of school-year basketball. I could already jump high enough to touch the rim. That was a big thing back then, which became very important for my game. I played for the school's team, playing against the Higgerson and Elliot Junior High. We won two tournaments. School in general wasn't interesting, so I didn't do well. I was working at a drugstore on the corner of Bower Street and Humboldt Avenue. It was called Chinies. It didn't have a soda jerk, as it was called; however, it did have its benefits: free ice cream to take home, free candy, and what became very important—free cigarettes and liquor (of course, what I wasn't supposed to have). I had my working permit. This wasn't hard to get back then because it was expected that black boys would quit school to work

At some time during that year, I began my first adult lesson in life. My cousin Joan had a girlfriend, whom I won't name, who lived across the street from my school. I can't remember how it started, but I do remember like it was yesterday what happened. I was introduced to the sexual world of manhood, starting off and lasting about fifty-one seconds. However, eager to learn, I continued to be trained almost every day, learning all there was to learn about satisfying a lady, including the fifteen pressure points on a lady's foot that they like to be massaged. You understand what I am talking about? Me and my demon were having a great time. I now wonder if this would make me eligible to move. Not that I would join. After a year, I got my PhD. In order to cover my tracks, I had to tell my school principal Mr. Guidon that I had to leave school to work in order to help my grandma. He went for it and suggested that because I only had one more year to go, perhaps I could come in part-time. I quickly agreed—in part because I think he didn't want to see my grandmother again.

CHAPTER 6

Getting Acquainted with My Stepmother

The following year, my father got married to my stepmother Ethel, who was and still is a very beautiful woman. I want to say right now that I wrongly mistreated her affection that she has always given me. My growing-up problems that she had nothing to do with deeply affected my relationship with her for many years. I'm deeply grateful for the chance to try and make up for all the trouble I have caused her. She will be one hundred years old this year and is still going strong. It took many years for me to adjust and appreciate the life my father and my stepmother tried to provide for me. I guess this is where living and learning steps in.

They were married at St. Mark's Church on Townsend Street, Roxbury. Later that year, my grandma let me go to Oklahoma to visit for two weeks. I have to admit, it was a fun time. I rode horses and drove a horse-team delivery cart that brought milk up to the state hospital in Taft where my father was in training to become a psychiatrist. My stepmother let me drive her Pontiac to Muskogee, Oklahoma, about thirty miles from the hospital where she lived. She said she did it so I wouldn't borrow someone else's car. I even went to a Western rodeo—the real thing. To tell you the truth, I don't think I had time to think about my problems. However, all good things come to an end, and after I was home for what seemed like a day, everything went back to normal. However, I did start renting a horse team to go around and pick up rags, old newspapers, and any medal pieces I picked up. I sold them to an outpost across the street from the now known Goodwill store. Let me add that my brother and mother don't believe that I drove a team here in Boston.

During that time, there were people who earned money by doing this weekly. I gathered these things and sold them to the horse barn.

CHAPTER 7

Living with My Parents

The following year, things changed again. My father and stepmother had moved to Tuskegee, Alabama, where my father worked at the Tuskegee Hospital Institution. My grandma was once again trying to get us together, so this time she pleaded with me to try again. So what could I do? My grandma and I drove there from Boston, having no problems until after we got south of Washington, DC. Then things changed. I was hit with my first revelation of what real racism was. My father pulled into a gas station for gas, and I had to use the bathroom. I didn't see the sign saying "Whites Only," so I used it. When I came out, everyone was looking at me. My father apologized for me not knowing, and we got into the car and sped off. After we got a few miles, my father pulled over and started yelling at me, saying, "Don't ever do that again!" I didn't understand what I had done. When he came up for air, my grandma explained things to both of us, telling my father that he must understand that I had never been down South, then taking time to tell me how things were as she had often told me about.

So as we went on, things were further explained to me. The more I heard, the more I was beginning to think that this was no place for me. However, I had no escape plans. We got there, and my grandma soon left after seeing how things were going to be. Of course, I felt like I had been betrayed.

I guess this was as good a place as any to try and get you to understand what was growing inside me. I call it my demon. A kind of under-the-surface anger that would always be there, ready to attack anyone at anytime. I couldn't be controlled. Family love and affection meant nothing to me. All I wanted was to be left alone. My grandma was the only controlling influence on me, and that wasn't always working. I was beginning to be someone else, especially when I got mad or felt restricted. I started to become numb to the pain. Didn't want to be close friends with anyone. All these feelings made it impossible for anyone to provide me with a home, even at this early age. All these feelings affected my father and stepmother's ability to do anything for me. The one positive about all this was, I never became a criminal. I am not making excuses for my actions. I don't think I have to. I just want to point out my state of mind. Plus, this is one of those rare opportunities to express your point of view uninterrupted. Thank God I never did any real harm to anyone.

My living conditions were great. My newly born brother lived there also. I will talk more about him and my other brother and two sisters much later. I was enrolled into Tuskegee Institute High School where it wasn't long before I got to meet and become friends with some boys that would always be singing outside in the back of the school. In class, I started having trouble because the studies for my grade was a year behind, so I was bored and really didn't pay attention; but when called upon to answer, I always had the right answer. I started playing basketball. I didn't know that one of the players was considered a star. When I played against him, I guess I made him look bad. We got into a fight were I beat him up pretty badly—not because I was better but because

I had more pent-up anger in me, so my body couldn't feel anything. The demon in me had me.

My stepmother was called to school and straightened everything out, and nothing else happened. Even the boy I was in the fight with forgot all about it. I felt badly for having taken my anger out on him. However, that was the way most fights started and ended in those days. The strange thing was, my stepmother never told my father. At that time, I couldn't accept it as I should have. Overall, I guess you could say everything went well at home. I didn't enjoy going out anywhere with all the restrictions. In town in the movie house, we had to sit in the back where the floors were dirty. In Montgomery, they were just starting the bus boycott.

I had to almost always, it seemed, to have to stay in the car because my father always thought I would get into trouble. In that case, he could have been right because I was very curious about everything. I did come to realize that this was the same racial problems in Boston, but it wasn't in full view. As time went on, I came to see it and found myself deeply involved in fighting against it.

There was one adventure I can remember. It was sneaking out of the house and finding my way to the so-called Hoochie Coochie Club back in the woods were all the black folks went to party. I was told about it at school, and you know I had to see it. Wow, what a treat, hearing all that down-home music. It was something that I have never forgotten.

I clearly remember seeing the Tuskegee Institute Tigers play Football down a big hill. I also remember seeing MLK speak at the institute chapel the Sunday before lighting struck the church and burned it down. To be honest, looking back, I believe I missed out on a great opportunity to begin having a great family life. But as they say, "Hindsight is always best."

Finally after a year, we were moving back to Boston. You know my brain started working on a possible plan of escape.

Soon after returning to Roxbury, my father and mother moved into a big white house (number 12) on Seaver Street on the corner of Seaver Street and Walnut Avenue. I was enrolled into Jamaica Plain High School, which meant discipline, and you know that wasn't going to happen; so right away, I began making my next move.

CHAPTER 8

Taking a Big Step

I started working at Filene's in Boston, running the freight elevator on the dock. I can't remember how I was able to do it. But it was a good-paying job, as jobs went for black employees. I was making as much, and with overtime, that meant you worked forty-four hours first. Especially during sales, I made more that my aunt Bertina, who had been working there for over ten years by this time. Remember, I had already gotten my working permit. I was working on the loading dock. This allowed me to get a lot of ladies' clothing that were being stored for sales. Of course, you know I took advantage of this opportunity to make a dollar. I looked at this opportunity as a gift. Well, that's how I viewed it at that time.

I even got my own apartment from, of all people, Ed Brooks, the Massachusetts district attorney at that time. He was the state's attorney general during the Boston Strangler case and later became the first black United States senator.

I rented the basement apartment on 26 Crawford Street for about $30 a month. I was able to sell the newly worn pantyhose that the young ladies wore for their dancing at Elma Lewis Dance Studio in Grove Hall, Dorchester. Also the new seamless hosiery with the garter strap. They would always come by for at least two pairs. Of course, I received fringe benefits. But I never mix business with pleasure. I sold every woman's garment that you could think of, so you know business was great. I had a phone, television, and was cooking.

My grandma, realizing that I had my mind made up, didn't like it but understood. I still came by often to help her. I kept a lot of my new clothes there. This cost me during that year because Herald stole all my new clothes and ran off with a girl that I had been seeing and that was babysitting for them. He never returned, and Adelaide finally declared him dead. That was something that I didn't see coming. For a while, as usual, I was more to blame than him for doing it. You know how that goes. But I didn't let it bother me because I could go home. From time to time, I even saw my teacher. Oh, how things had changed.

My parents continued to harass me to give up my apartment and even took me to court. However, I had a lawyer who got the judge to understand that I was supporting myself, so he said based under these circumstances, he couldn't do anything about it.

That year, my cousin Richert bought a car, an Oldsmobile 98, a powerful car that had SuperDrive. This was a new feature. So you know me and my devil. I got him to let me drive. I didn't even have my driver's license. Letting me behind the wheel was his big mistake. So you know what happened. I got my friend that lived next door, and I drove all the way to my aunt's house in Washington, DC, before my cousin and neighbor could do anything about it. I don't know what got into me, but I couldn't help myself. I guess you can say it was another case of "the devil made me do it." When we got there, we sat outside deciding who would ring the bell. Of course, you know it was me; they were scared to move.

When we came in, my aunt immediately said that she had received a call, and they knew I might show up. "So turn around and take your behind straight back home." I said we needed gas and toll money. She said follow her to the gas station, where she filled up our car, gave me $7 (remember, gas was only $0.27 cents per gallon), and then she said, "Don't get pulled over, and I am calling to tell them you're coming." I had time on my way home to think about my next move because I was still the driver, partly because my cousin had never driven on the highway, and I had. By the time we got home, everyone was sitting on their front porch waiting like a sheriff's posse. So I dropped my friend off on the corner and took off for about a week until things calmed down. My cousin just went home with his car, saying, "You always cause trouble."

I went to Nantucket Island to work. I got a job running the telephone switchboard and picking up and dropping off Breakers hotel vacationers. This was a famous hotel. I got to meet some famous movie stars like Frank Sinatra, Dean Martin, Sammy Davis Jr., Abbott and Lou Costello. I also met lady movie stars and other important celebrities. They came with their yachts. I made my money mostly by being available and not being seen. You know, "see no evil, know no evil, and speak no evil"—just like my grandma taught me.

By the time I came home, I had almost five hundred dollars, in part because I didn't have to pay for anything while there; I only had to be constantly available. When I got home, my grandma, of course, thought I had robbed a bank, but I convinced her by looking her in the face and asked her if I would lie to her about something like this. I may not have always done the right thing, but she always told me to stand up and take my punishment. She then asked me, "What are you going to do now?" I told her I didn't know, but something would come up. She smiled with that certain smile she had and pulled me close and rubbed the top of my head, something that she always did to show her love. I said, "I want you to hold this money." What was I going to do with so much money? It was like giving a ten-year-old a real plane. Plus, giving it to her, I knew it would be safe. She and I had come to realize that I had a lot of the wandering devil in me, but I was basically a good boy. She smiled again and said, "That's my boy," and we played cards and checkers, which she still beat me at.

It was nice because my aunt had moved, and my uncle wasn't there. However, my cousin Joan still lived there. Now because I continued to be harassed by my parents, I decided that I would try and join the air force.

CHAPTER 9

Trying Out the Air Force

Here I was, fifteen years old but tall for my age; plus, I had some convincing ability to get past problems. I was sure I could do it, so I began my new plan. The recruiters' office was in Dudley Street right where I had always gone to pay my grandma's and my own phone and gas bills. So I started standing outside getting a feel of what was needed. It wasn't long before I was ready for my next move.

Being the person I was, never satisfied or scared to try almost anything, I had my cousin Freddie's birth certificate. He was old enough to sign himself into the service. I went and passed the test and then went to Charlestown Navy Yard to be sworn in and take my physical and passed. I was told that I would be leaving in a few days, so I had to move all my things out of my apartment and let Attorney Ed Brooke know that I was giving my apartment up without telling him where or what I was doing. I stayed at my uncle Charles's house for the next couple of days because he never got involved in others' affairs. When it was time, I took a deep breath and left.

I was soon at Logan International Airport, waiting for my plane to go to San Antonio, Texas, with my eyes popping out of my head. When I arrived, I was picked up along with some other recruits and taken to Lackland Air Force Base. I couldn't believe I had made it. I went through the processing and got my haircut, clothing, boots, toilet supplies, including saving gear that I was many years away from using. We went to our barracks and was introduced to our drill sergeant. We were taken outside to immediately learn how to drill, salute, and the rest. I didn't mind getting up at 5:00 a.m. because I had always done so. In fact, I was already up before morning call. I lasted about a month when I was called in to the office. I knew the jig was up, so I had little time to figure out my next move. I went in and saluted and was told to sit and wait. As I waited, some MPs came into the office, and I just knew they came for me because they were waiting also. Thank God they went in and left without looking at me. Now it was my turn. The captain in charge said, "What do you have to say for yourself?" I remembered my grandma telling me, "When you are in trouble, always tell the truth." So I did—in a way. I said that I was supporting myself and was getting into trouble, so my only way out was to join the service. He looked at me and said, "Son, that's the best story I ever heard. So I'll tell you what's going to happen. You are going home. So get your things packed and don't come back until you can use your own name." I saluted and left and had to immediately go to the bathroom before I peed on myself.

I got my release papers and enough money to get home by any way of my choosing. So I decided to take Greyhound. That way, I could save a few dollars and stop in Tuskegee; plus, I was in no rush to get home.

I took the long ride to Tuskegee, not remembering that I was in the South and was restricted to the rear of the bus and couldn't use the one bathroom on the bus. I had to wait until there was a bus stop and then had to use the dirty "Blacks Only" bathroom. It was like God was punishing me or teaching me a future lesson.

I did finally make it to Tuskegee but was constipated from not being able to relieve myself. I quickly learned that I wasn't as knowledgeable as I thought. I soon got myself together and enjoyed myself for a few days, spending time with some of my former friends at one of their homes, finally getting to go to the Hoochie Coochie bar. Basically because I still had my uniform on, I guess I had stretched the truth a little—to the point where I was loaned the use of a doctor's car, who recalled me living there with my family for about a year before. I told him that I had joined the air force and was returning home for a week or so. Not quite the truth but close. I ended up having a fender bender but couldn't pay for it. My parents did. Just want to tell it straight. Anyway, no harm done. After a couple of days of fun and having a moment to enjoy the young ladies, who knew that I was full of experience, upon arriving home, I soon found out that it was my stepmother who found out where I was.

I immediately got my job back at Filene's. This time, I ran the press room—this was even more money than before—bailing two hundred pounds of paper that would come down from the ceiling floors. Don't remember how or why I was taken aback. I would like to think it was because I was a good employee. I even got my old apartment back. It was like, in a way, nothing had changed—especially because I was back in business. You know what I mean.

It was impossible for me to get my parents to understand that I was someone that couldn't be tamed. I was controlled by my wanting to make up my own mind about my life, in part because I had survived this far without their help. So eventually, they gave up trying. Now looking back, I feel guilty for a lot of things I put them through. However, it wasn't that my father wasn't involved; and if he had used his newly found psychiatric treatment, by examining my problem, he might have understood his own involvement. I tried for a while to adjust to being home, but for some reason, I got restless again.

CHAPTER 10

My Second Time in the Air Force

I was being harassed by my parents, and I had already appeared before a judge that determined that because I was supporting myself and his hands were tied, there was nothing he could do. My grandma was pretty happy that I had returned, but I was getting restless again. I felt the devil in me calling me again. So this brought on the second major change in my life. I went back down to Dudley Street, and this time, I had my own birth certificate where I, of course, forged my age and passed the exam. Not the same airmen were there; I checked first. I had to be on my game. I had already told my job that I was going on active service, showing off my papers to go to San Antonio, Texas.

I told my school that I was moving, which was technically true. I left Boston for active duty in late 1958. Boy, was I excited. I knew I had it made this time. I arrived at Lackland Air Force Base, went through the admission process, and was lining up with my group when the drill sergeant asked if anyone had any prior military experience. I immediately said I did before I knew what I was doing. Thank God he never questioned me about it. I believe it was how I stepped forward at attention, saying loudly, "I do, Sergeant," remembering my past experience. When we went to our new barracks, the sergeant said that I would be the barracks chief and was responsible for the way the barracks looked for inspection. I knew I had put my foot in my mouth but knew it was too late to back away. So the first thing I realized I had to do was to get all these white boys to follow me.

I was the only black boy there, so I immediately called everyone together and confronted them with the situation, saying in general that we could all sink or swim together. If we couldn't agree, we would never get a pass, and they would never learn how to keep their bunk and footlocker straight. After some grumbling, they agreed and fell in line.

Once again, I got through a tough situation. Immediately they learned how to do things—like how to make their beds, folding the sheet back to the length of a dollar bill, folding their underwear to the merger a dollar, spit-shining shoes, which I, of course, knew but found it funny seeing all these white boys never having to do it having so much trouble. And of course, there was latrine cleaning, which I always assigned to those that I had trouble with the general upkeep for the barracks, checking with a white cloth that everything was spick-and-span.

The drill sergeant was very surprised to see how well we were doing after our first inspection; I don't think we had more than one demerit against us. With that, the white boys were my best friends; plus, we received our first off-the base pass not long after. This was my first experience seeing the ladies in San Antonio. After being there for about two months, I was called to the base office. I just knew I had been caught again, trying before going in to think of an excuse. I went in and gave the proper salute. My captain said that I had orders to go to Germany. Needless to say, I almost fell over and asked him to repeat that. He smiled and said, "Son, I guess you hit the lottery." I was told to pack my things, report to transportation, and I would be carrying my orders starting tomorrow morning. I saluted and left with my feet not touching the ground.

I started to have a conversation with myself, asking what just happened. Is this for real? What did I do to deserve this? By the time I got back to my barracks, the men already knew; my sergeant had told them and that I would pick the new barracks chief. That night, we were allowed to have a barracks party (no liquor). That night, I made my choice, no problems. I packed that night; then the next morning, while leaving, I, for the first time in many years, felt bad leaving my new friends—and they were all white.

Now can you beat this? I went over to transportation where my orders were waiting, along with a Jeep to take me to the airport. Now keep in mind I didn't even finish base training yet, not even one strip. Here I was on my way to Germany. When I got to the airport, I had some time to waste, so I opened my orders to see if this was for real. Well, was I surprised—everything was correct, except the last name. They had Kenneth E. Waldron when my name was Kenneth E. Walden. ID tags were correct. So now it was decision-making time. Needless to say, being prone to doing something new with the help of my devil, I said, "Let's do this." So I became Waldron. I figured not my fault; I was just following orders.

I was given a one-week leave before going overseas, so I hid out in Boston because that's where I was leaving. I made it through that week and can't remember how, just that I went to the Charlestown Navy Yard to get my pay and airline ticket to fly commercial airline to Frankfurt, Germany. I still couldn't believe I was doing this.

CHAPTER 11

Going to Germany

In Boston international terminal, sitting with the bigwigs (white folks) waiting to board my TWA airplane going to Frankfurt, Germany, I felt like someone important. A strange feeling for a black man, especially for me being underage. I still felt insecure, thinking at any moment my dream would become a nightmare.

Now boarding, I of course had to sit in coach. For me, it was first class. Plus, being in uniform, I felt like I was someone important, carrying my own orders. When the plane took off and I looked down, I remember thinking, Goodbye, Boston, I'm on my way to a great life. We had two stops: first, Newfoundland; then Shannon Ireland; then Frankfurt, Germany. I can't remember how long the trip took, but I didn't care either. Once I landed, a Jeep drive with two strips was waiting to take me to Rhein-Main Air Base to be processed.

I recall that here was a two-striped airman driving a no-stripe recruit. How strange this is. I thought he was going to salute me. Going into the processing barracks, I thought this was where it would all end, but things went well, and I was told that I would be sent to Munich in a couple of days. In the meantime, I could have the freedom of the base. like chow hall, movies, and recreation facility. I was given a place to sleep, and I started wondering when would this joyride end. A couple of days later, I was called in and given my ticket and traveling orders to go to Munich by first-class train. Once again, I was spared.

At Hauptbahnhof German Train Station, I caught the Zug train and began to learn a language that I didn't understand. I was being looked at like I had two heads. However, I was treated well and started learning to eat German food. But I refused beer, which seemed to be drunk by everyone, including children. I was very observant, learning from my grandma that you learn more by keeping quiet than by opening your mouth and letting people know how ignorant you are. So you see, the devil wasn't completely in charge.

I learned quickly that German trains leave on time to the very second. I traveled mostly by day and remember passing Stuttgart, Germany. I noticed that the train was traveling higher than planes landing at the airport. It was amazing to be able to see the people in the plane so close flying below me. I started to really think that I have much to learn if I was going to survive here. I finally got to Munich. I remember it was late at night, and I was being picked up again. I stayed at the army base there in Munich; I think the name was Kelly Army Base. I do remember that it was a big processing center with the single-person elevator service where you got on by yourself, and it moved at a pace where you could get off. I think that stood out in my mind because of my time driving one at Filene's when I transported the bales of paper.

The next morning after chow, I was picked up and given a train ticket for Freising Air Force Base some forty miles outside of Munich. I arrived at the Munich transportation at about 9:00 a.m., and the train left at 9:21 a.m. exactly. Put that in my memory. About an hour later, I arrived at the

Freising Station and was picked up by another two-striped airman and driven to the base, which I remember was up a big hill. There were German guards at the gate, which concerned me. I was driven right to headquarters and began processing with no problems. I handed in my orders; and of course, the names didn't match. But my service number did, so they said that it was another of the processing center's mistake. I was greeted by the base commander. I believe he was Lieutenant Colonel Galvin. Then I was told I was a replacement for the motor pool.

He said take a couple of days to get my bearing and then report for duty. I saluted and left, beginning to feel secure. Next I was driven to my barracks, which was the home of the 604th AC and W squadron, where I met the other airmen, all in their late twenties and all at least technical sergeant and all black after the introduction. I was helped in getting settled, and they asked all kinds of questions about the States. I felt, for the first time, taking the place of my grandma, having all the information.

I think this exchange went on for days, even when I went with them for chow. I quickly learned that blacks weren't liked a lot, and I would have to learn much to survive around here without getting into trouble. So stay close, keep your mouth shut, and learn. It sounded like words I have heard before and knew that they were important. So you know I did. For the first time in years, I felt that I had to take advice and orders without question; for some reason, I found it very easy for me to do. I guess I had more smarts than I thought. A couple of days later, I put on my fatigues and brogans (boots) and reported to the motor pool. There I found myself standing out because they were all white, and I was taller than all of them—I was 6 feet 3 inches and weighed about 195 pounds.

I was asked if I was familiar with trucks. I said yes, quickly remembering my early years riding once in a while on the highway taking the wheel of my uncle's Mack truck. So I had the job of pulling the brake lines, taking the ball bearing rings, repacking them with heavy grease, and putting them back. And the real joy was driving them to and from the parking area. I didn't have a license, but that was soon taken care of without me taking a test. Back then, it wasn't hard to get a service truck driver's license. Remember, they thought I was at least eighteen old. After my first day, I was quickly accepted by the crew.

That evening, I was Invited to join them at the base's game room. Of course, I went, still keeping my mouth shut, which wasn't easy at first, but I quickly learned much by doing so. After a week or so, I had my first trip driving one on the Autobahn high-speed highway by myself following other trucks to Stuttgart, Germany, at least 250 miles from our base. I felt like I was in heaven—who could ask for anything more? We got there, delivered the trucks, and returned by train. I got extra money for every trip. I felt like my uncle.

I was given my first pass to go to Munich. My eyes were wide open and, of course, my mouth shut. I had received my first pay and immediately signed up for US saving bonds of $25 per month. Because everything I needed was free and anything I wanted, I quickly learned that I could sell my ration stamps, cigarettes, liquor in Munich for a good profit. I learned right away that the value of money or dollar was about four times the value of their marks, which was the German currency. So you see once again, I had found a way to make money. It was technically illegal, but it was commonly accepted. Even the MPs were doing it, so I didn't feel that I was doing anything wrong. I had to quickly find a way to get some civilian clothes because no one wore a uniform off base.

So that was my first order of business. Strangely enough, I had little trouble following orders. For the first time, I was being told what to do but had no trouble doing it as if I had a choice. After a month, I found my way around the base and learn that our base was shared by the German Air Force, with which we had no real contact. I went to the movies a lot during the week for twenty-five cents. I also found my way to the gym, which became my second home. I practiced every day and was soon asked to join in on intersquad basketball. This led to my joining the base basketball team.

While playing and showing my unusual skills, my hook shot and ability to rebound caught the attention of the playing coach, who asked me how I learned those skills. I said practice and watching the pros at home and pickup games. He then asked me if I wanted to try out for the base team. I said yes and asked when and where the team played. He said all over Germany, and the base commander was a basketball fan. So you know, that piqued my interest. Soon after, I was accepted and started practicing.

We had three blacks, which included the coach whose name was Sergeant Harris, and another black sergeant whose name was Glynn. I don't remember the white players, but we got along while playing together.

By the way, during this first couple of months, I got used to driving the dues and half trucks. It became like a bike. Plus, I quickly found out that selling my rations on the road where there was less competition and much less of a supply made the trip a more profitable experience. We stopped at places of the Autobahn. Nothing suited me better.

Going on basketball trips didn't provide much side money. I knew I would have to make a choice. I was tormented with the words my grandma said: "You can't have your cake and eat it too." I sure tried, but soon because the prime basketball season was beginning, I had to give up, for the most part, my truck driving. But on off times I continued to drive the trucks, telling my commander that I felt I should play my part because playing basketball, you were free of any other responsibilities. He complimented me on that and okayed it.

So now I feel I enjoyed the best of both worlds. I never caused any trouble doing my job and began to help my team win games, where I started being written about in the Southern area command's Stars and Stripes sports page, calling me "Junior Flip average 14 points a game." Blocking Shorts. Being taught that when I rebound a offences rebound tap it in and not come down with it. I learned much about the game that I didn't know and was able to get used to my advantage. Thank God I had enough since to keep my mouth shut and listen and learn.

As the year went on, I felt free and really began enjoying my off time in Munich at the clubs that blacks could go, especially when the jazz players came almost every weekend, much cheaper than if I was in the States. My German was pretty good enough to get around, and I was no longer being seen as a tourist. I even got myself a beautiful young German lady. Of course, she was a bit older than me but a long way from being old enough to be my aunt or mother. I think she was in her late twenties. She lived in Munich and came to every basketball game we played there. I was beginning to be recognized as a star, which didn't hurt—not bragging, just stating a fact because basketball was still pretty new in Germany at that time. Just wanted to be honest.

When the team went on the road, the players all had their own cars, so most of the time, I rode with one of the sergeants, sometimes taking my lady friend with me. If you are wondering why I said "lady friend" instead of "girlfriend," it's because I never mix business with pleasure I was smart enough to know that when you are in the service and have no guarantees, you shouldn't get too close. Plus, it was my way at home, so I found no reason to change. Also, there was no forgetting that I was black, and I always was aware that most Germans during this time pretended to like blacks for the benefit of it. It paid me well to keep my head on my shoulder, alert, not going by the head in my pants.

The devil wasn't winning this game. As time went by, I got bigger and stronger, applying more moves that I had learned. Plus, I must admit, I was enjoying the attention that our team was getting around the Southern Germany bases, which our commander loved. I never had to stand for inspection or barracks check, where the sergeants living there enjoyed the benefits. I now had a lot of civilian clothes, shoes, and a couple of great tailored suits. I loved this better than drinking or smoking. I, for some reason, never found an interest in it. When we went to our usual bar in Munich, the bartender got used to giving me Coke on the rocks; that was more expensive than alcoholic drinks because Coke was harder to get. We would sell cases of it every first of the month when we got our new rations cards. There was more money in this than there was in cigarettes and liquids, which also sold well.

As I look back, I remember one of my biggest thrills was getting my first stripe. My command said it was a little late; however, I quickly replied, "Sir, better late than never." He smiled and replied that I would go far with that attitude. I continued to do both driving and playing basketball. It was my joy; it gave me freedom, respect, a secure place to sleep, eat, and was teaching me to correct my most serious problem: learning to take orders without talking or disobeying them. That was one of the best lessons that I learned while in the service. I believe, looking back, if I hadn't gone into the service when I did, I wouldn't have turned out the way I have, not perfect but a decent man.

As the year went, our team started our tour, my first playing at other army and air force bases around the southern part of Germany.

I remember one important game we played. I got out there and amazed myself. Well, let me tell you it was either God or even the devil, but I couldn't be stopped. My team kept passing me the ball, telling me to shoot, and I couldn't miss. On defense, every time they tried to block me out, I would be fast enough to get inside where they had to foul me to get the rebound.

In Munich at the Kelly Base, we played against the 508 MPs. I don't recall if they were based at Kelly or not, but they were built like football players and played like it also. The first half was a brutal lesson for me in being popular and being targeted. They pushed me around as if I was their punching bag. At the end of the first half, we were down 7 points, and my ass was hurting. When we were in the locker room, my coach said, "Welcome to the big time. Are you a man or a mouse?" The rest of my team just looked at me and said, "Think of your training." I used all my skills that I have learned from our playing coach Sargent Harris. The end result: I scored 24 points, 13 rebounds, and we won by 12 points. They left with their heads down, but some came over to me and shook my hand and said they had never seen someone like me before and said the most important thing: I could take the beating like a man. My coach heard that and patted me on the back and said, "Now you have arrived." The game was written up in our Stars and Stripes Southern newspaper. I am sorry that I didn't keep the paper, but as I was young, things like that didn't seem important at that time. We went on to win the championship of that area that season. Of course, our base commander was very happy with the first of a few more to come.

Now that the season was over, I was given the choice of going on a ten-day leave that would allow me to go almost anywhere or spend my time on and around the base—a very unusual offer for a commander to present to me, but it was a sign of how he took to me, I guess. I asked if I could go back to driving trucks. He smiled and said something like, "Son, you must enjoy this life." I said yes without telling him how much.

We took the Autobahn to places like Bremerhaven and Robinson barracks near Stuttgart, so there I was, taking my place, being moved up in the line in transporting trucks up and down the Autobahn. By the way, I enjoyed going up into the sky tower similar to the one they have in Toronto, Canada. You could eat and watch jets and other planes fly below you at a distance landing. I always enjoyed that view. Of course, I sold my supplies, was always about the Benjamins, even back then. I learned early in life that if you don't take care of yourself, then don't expect anyone else to. Of course, this had been going on for years. Plus, it offered assistance to the German people, and then of course, it allowed me to go to jazz concerts in places like Copenhagen, Bremen, and Stuttgart.

Now you tell me, would you rather be anyplace else? Only if you were crazy, and I was not. I became very comfortable traveling on the train alone, speaking enough German to get around, knowing about money, asking directions, being able to read people so you didn't have to be taken, and most importantly, knowing when and where to go where I would be safe.

I was becoming recognized, but I was black, and you didn't ever forget it. I sharpened my instincts, which served me well throughout my life, as you will see. I can tell you now there is a lot to be said about your instincts. We all have them but mostly never use them. Things went well. I was starting to feel at home, really thinking about anything at home except for missing my grandma.

In 1959, I was free to drive trucks or travel, so I did both but couldn't help but think from time to time is, this the way service life was all about?

Later that year, I found out the base was in the process of changing to a German base, so there wasn't much of the ordinary duties to be done, except working at the radar site and motor pool. Most of the men, including the sergeants that I first knew, had transferred back to the States for either discharge or reassignment. So I guess I had picked the right time to be there when the air force made their mistake.

In 1959, I got my GED high school equivalent. This, later in my life, became very important. To this day, I have no idea why I did it. I must say I was very happy to be back with the same guys. This made me realize that I had missed the relationship and being around the right people, which wasn't so bad.

We practiced hard to check out our plays. Plus, since I had a full season under my belt, the player and I were much more comfortable, knowing where each of us was as the game went on. There were no new players available because the base was continuing to shut down.

This was the beginning of our 1959/1960 season. I was now seventeen years old and stood 6 feet and 4 inches, weighing around 205 pounds and wearing a size 12 shoe, which was big at that time. I had saved my money from my truck adventure, so I didn't miss it so much. That doesn't mean I didn't miss the driving part. We started playing during Germany's Oktoberfest. I never figured out what the meaning was, but I do know it was as important as our Fourth of July. We had a new schedule playing at some places like Stuttgart, Rhein-Main, Bremen, which was very far from our post. We looked forward to that trip, along with our usual

Kelly Base schedule. The most important game I remember was our usual game against the 508 MPs, at which, I am proud to say, we beat them twice during the season and in our playoffs leading to our second straight championship.

Everywhere we played, we had a pretty good crowd of service members watching. The only other time that there was another better known than us was the Harlem Globetrotters, which visited every year. We brought home four trophies. I now felt comfortable that I was old enough to have enlisted on my own. During that summer, more of our service members were leaving. Less truck trips were happening, so more off time was spent taking the train to the cities to tour around and see the sites. I had a permanent pass that allowed me to travel anywhere in Germany as long as I was off duty. You couldn't ask for anything better. Not that I was gone weeks at a time. I still had to be on base to get my pay. Remembering that when I first arrived, I had signed up for the $25 a month saving bonds. By now I had more than twelve, remembering my grandma had taught me that if you can save your pennies, the dollars will take care of themselves. I began spending a lot of time in our base library, not much but more than I was used to. Mostly geographic magazines. I enjoyed spending time with this mostly because it was history, which I loved and found that it verified some of the questions I had in school. Along with this, I continued to spend a lot of time in the gym, sharpening my skills.

For some reason, I lost interest in entertaining the ladies but had more time to enjoy the jazz concerts that appeared to be more often with more artists. The summer passed with me growing not just in size but in my state of mind. I guess you could call it maturity. I think it served as an example of "when you are a child, you play with toys; but when you become a man, you put them away." Something like that. I'm trying to remember what my grandma said. I guess you can see how much I loved, missed, and respected my grandma. She raised me the right way, leaving me to make my own decisions but remembering that there is always a price to pay.

I didn't always follow her teaching, but I did always remember her telling me what the right thing to do was. By the fall of 1960, we were told it would be our last season together because the entire base was being turned over to Germany. So none of us were very happy about it but felt that this should be our best year.

This year was the most important period of my time in the service. I didn't know it at first, but as the fall season got going, things started to change. During our first trip, one of our key players got hurt; another got called for return to the States for discharge. This created a problem for us that wasn't easy to overcome. We still were winning, but their absence caused all of us, especially me, to not feel the same way. The three of us—Sergeant Harris, Sergeant Glynn, and myself—and two-striped air force members were pulled to join an all-star team to play in Bremen, Germany, some seven hundred miles from where we were. I had been there once before driving trucks. I knew if I was chosen, I must be known.

We traveled by train with first-class accommodation. Three black men riding first class in Germany at that time was a sight to behold. It was like a white man driving a black man around in Alabama in the 1950s. We arrived and joined others that I recognized from the various teams we had played. We played teams from England for a week, winning four out of six games. It was a kind of tournament. I didn't have a chance to look the city over; it seems to be what happens when you go places in the service. You go places, but you don't get a chance to tour the city if you're black—at least it seemed that way at that time. A case of being entertainment for the white man but not good enough to be able to join them. I think because I was young and wasn't used to it, I took it to heart but had enough sense not to rock the boat. We returned and joined our team and continued to play. However, this was when my trouble began.

My commander called me and asked me to close the door. I knew right then and there that the jig was up. He said he had received a report from Lackland Air Force Base that they had made a mistake regarding my orders. I thought of what my next move was going to be but quickly decided that the whole truth was best, so I agreed and told him I was aware and knew that they had made a mistake even when they had the correct ID number. After some discussion, he surprised me by saying in a strong voice, didn't we have one last championship to win? I surprisingly said, "Yes, sir!" He ended the conversation by saying I had been here long enough to be able to sign myself into the service, and the orders didn't include sending me back home, so we would keep this to ourselves. I tried not to smile but saluted and said, "Thank you, sir." I was once again going to the edge but not crossing over.

That winter, something else happened that could have changed my life forever. I was in Munich shopping for myself and wasn't aware that our team was asked at the last minute to travel to England for a Christmas benefit game. That night, I believe it was the seventeenth of December, there was a serious snowstorm, and the plane that we should have been on crashed on the main street in Munich not far from where I was. All aboard and many on the street were killed (Appendix 4 will show details). My team and my commander called me their good-luck charm. I don't know about that; I was too scared. It took a couple of weeks for me to get over it. However, we did go to Robinson army base in Stuttgart to win our last championship.

Now it was time to go home. In 1961, trouble was brewing up in Berlin, where we had played basketball several times. My first decision to make was, what to do with all the things I had purchased? I couldn't take them home: stereo grudge radio, suits, shoes, shirts, and other things not my watches and other gold jewelry. So what I did was have a, like what you call today, a garage sale. That was well attended by the German troops on base. Didn't do too bad. Nothing is better than money when you are going home, something I had learned well.

There was a going-home party for the team, where the base commander appeared—not normal for a white officer to come to a black club in Munich. But it was a great surprise and a true sign of respect. This was one man I truly respected. After a short time, he left but said with a smile, "Thanks to you men I am going home with the most winning basketball trophies in Germany." We got to attention and saluted. All of us really appreciated that moment. The last time I had a chance to speak to him was when I went to the administration barracks to pick up my traveling orders and train and plane ticket to the States. He wished me well and said, "Don't worry about getting into trouble." He said that he had written a good report about me and that the service would be well represented with me in it. I shook his hand then saluted and left. I packed my duffle bags, two with uniforms and clothes that I kept, and went to the motor pool to say goodbye to the few men left. One of them drove me to the train station in Freising. I said goodbye, and the train came and left on time.

Now I realized that I couldn't take time dreaming about the past. I had to make plans on what I was going to do next—like what was I going to face when I got off the plane? Would I be arrested, court-martialed even if my commander had given me a good report? Would it even matter? I was on the edge once again. But in the meantime, I told myself that I should enjoy the moment because whatever was going to happen was out of my control. Plus, it wouldn't be my first time being at the cliff's edge.

I rode first class to Frankfurt, Germany, and was picked up and driven to Rhein-Main Air Base once again for final processing. Much to my surprise, it went well—no upsetting questions, very normal interview. I was even told that I was lucky going home because trouble was brewing in Berlin. I didn't pay it any attention at that time. If I had, I might have talked my way into staying overseas. But within two days, I was at the airport flying home on a MATA (military air transport aircraft), full of returning servicemen all talking at once, showing pictures, talking about what they were going to do first when they got home. Most of the time, they talked about facing their wives and girlfriends, wondering mostly if they were waiting for them. I thought that was funny when it was a fact that most of them hadn't cared enough about their ladies while they were having fun here. However, I pretended that I was tired and needed sleep and was left pretty much alone.

We stopped at the same places that I had coming over, but the big surprise was that we landed at McGuire Air Force Base in New Jersey. I reported to the discharge barracks, thinking, Here it comes. But much to my surprise, I didn't get any of the problems I had thought was coming. I was asked if I wanted to re-up, and I would get my second stripe. I asked, did that mean that I would return overseas? He smiled and said, "You can forget that." I thought for a moment all the extra money I had been getting for overseas pay and all the travel money I had been getting.

There I was, facing another white two-striped airman and little bit jealous and would be sending me to some places like Thule, Greenland. You see, I had done some checking before coming home. So I took my honorable discharge and went home. Wouldn't you know it—twenty-four hours after

I left the base, the government stopped all discharges because the Berlin Wall was going up. I said to myself, if I had stayed in Germany for two more days, I would've been still there. But the good part was, if I had still been at McGuire Air Force Base, I would have not been able to be discharged, and that would mean possibly being sent to Thule, Greenland. I had to look at the bright side: I was going home.

CHAPTER 12

Coming Home, Pain, Tragedy, and My Greatest Loss

So here I was—coming home, feeling strange, beginning not to identify with my surroundings, and having more money in my possession than I ever thought I would ever have.

Immediately after arriving at Logan Airport, I felt like I was in a time warp. I was 6 feet 4 inches and weighed 210 pounds. I was in great shape but no plans. I guess that's how most service members feel when they come home, especially from overseas. I guess the education that you receive has much to do with your growth. I know in my case I felt much differently now than when I left.

To tell the truth, I guess I am saying that I had left as a boy thinking I knew it all and returned as a man—disciplined, educated, much more observant, and in general grown up and ready to take on much more responsibility. If it hadn't been for the service, I believe I wouldn't have made it.

Everything changed when I was told my grandma was dying in Saint Elizabeth Hospital of polio. I immediately went to the hospital, and when I entered her room, most of the family members were there, surprised to see me. I went up to her bed where she was balled up, not moving. So I put my hand on the side of her face and leaned over and said, "I'm here, Grandma." She rolled over lifted her hand, rubbed my head as she always did when she was pleased with me, and said in a soft voice, "I've been waiting for you. Remember what I have told you, and God will show you the way." Then she smiled and died.

That was the worst moment of my life. I looked up and could vaguely see the jealousy on my relatives' face.

When we all had returned to my grandma's house, my family started right away deciding what each one was entitled to. I got so upset that I made everyone leave. I spent that night just sitting and remembering everything I had experienced with my grandmother, feeling her presence. It was like she knew this day was coming and was prepared. I know this sounds crazy, but this how close me and my grandmother had always been, even today. I always talked to her, especially when I needed her advice. The next morning, I had calmed down and just wanted to know what had happened. I was told that the house had been sold under "eminent domain." I asked, what the hell was that? They said it was a new law that allowed the state to take land owned by someone, give them what they felt was a fair price, and have them leave within a specified time in order to provide new housing that we will be the first to return to. I went into shock, saying, "You believe that bullshit?" Because I knew my grandma would never have gone for this type of statement. I could hear her answer: Why do you need a new housing when the houses we have now are in perfect condition? Plus, my grandma lived by that old rule: "Don't believe anything that the white man say he wants to give you, especially when he has his nigger with him." I could only say, "Is this what caused Grandma to die quicker than she had to?"

She saw no way out, fighting as long and as hard as she could without much help. Everyone just looked and said nothing—I believe because they knew trouble was on the way.

All of my demons came forward. I could feel my blood boiling, and I saw red. I wanted to go to the state house and get back all that they had taken from my grandma, knowing that it wouldn't bring her back. What was even worse, this new disease called eminent domain was the cause of it, taking only the homes of black families who all came here from the late 1880s during Reconstruction, immigrants who worked like hell to uplift themselves. We were being targeted for reasons I wasn't sure of at that time. But I knew it wasn't good. My grandmother had just paid the price, and I wasn't there to support her—a guilt I carry even today.

However, I wasn't a fool. I knew that I couldn't go back and change the past but felt that was part of what my grandma was telling me in the hospital; at least that's what I believed. There was much more to this than what I was being told.

Here I was, coming home, thinking I had no plans, and maybe as my grandma had just said, "God will show you." I knew for a long time to never question my grandma. However, first things first. There was her funeral to attend to. She was buried at Mount Hope Cemetery in Mattapan. Everyone in the area attended; it was more than the church could handle. Everyone told me that she had missed me and said I would be home soon. It made me break down more than once. After the funeral, everyone, including my family, came to our house, eating a lot of food, singing, and talking about her life. Just as I had remembered my grandfather's death and how my grandma handled it, I tried my best to be the same way, but it wasn't easy. It seemed like it lasted for days. I even thought of telling them like I told the preacher years earlier that they should go and take their food else-where. But I had learned much having been in the service, and I guess this was the beginning of testing what I had learn, remembering what I was taught. Everything happens for a reason.

My family began splitting up her things. All I took was the radio that was in the dining room by the radiator, where we would always spend so much time listening to our favorite programs while I read her Bible. I couldn't even tell you who got the money or even the amount of money was paid for the house and how it was split up. Much to my amazement, my grandma still had hidden my money in her secret spot that I only knew about. She still had all the money I had given her before I went overseas. I believe it was even the same bills.

After taking a few days to get my bearing, it was time for me to find out the truth.

I learned that in order to argue your point, you have to have the facts. That wouldn't be easy because as I had just found out from my own family. They really didn't want to challenge or know the facts. Just agree and take the money. Well, I don't need to tell you that I was raised to challenge everything and stand up straight for what you believe in. Even if you lose, it will be well known what you thought. So with that in mind, I began my quest.

I knew from experience that the best person to talk to about the problems affecting black people was the minister. So I went to set up an appointment with him. He was a new preacher, as I understood. I got the feeling that I was being put off because, first, he was going to meet with me later that day. But later, I was told he couldn't because something had come up, and he would let me know when he could see me. I said okay but observed that the lady didn't ask for my telephone number or where to reach me. As I have said, this was the new and improved Kenneth Walden. I did leave, but

I had a plan to come to church the next Sunday and confront him, like my grandma confronted the school principal. When you pay attention and are taught the right way, you know how to get things done—and I was taught by the best.

The next few days, I was able to look over the papers that the state used to acquire the property. It was signed by Mayor Collins, Ritchert LOB (a land developer), and Robert Coard (director of ABCD, a new antipoverty program that was supposed to protect black people's interest). You know that immediately raised my eyebrows because it only reconfirmed my thoughts, and I could hear my grandma saying, "Always be suspicious of any offer given to you by a white man supported by his nigger." Little did I know she was right—again.

I tried to ask more questions about these people but couldn't get the answers that I wanted. That told me that a certain amount of fear was running around. That meant that getting the correct answer wasn't going to be easy, and I would have to watch the eyes of any person providing the answer. The following Sunday, I went to church and sat through services. Then right after, as the minister was greeting people at the door, I approached him and stuck out my hand, held his, and introduced myself, saying we had to talk here or later about this eminent-domain problem. I was still holding his hand firmly. He looked at me and agreed and gave me a day and time, which I had him write down for me just so I would have proof if I needed it. You can't trust people when it comes to business. I may be arrested for trespassing or something.

I learned to think before you leap and know where and how far and high you want to go. All these things I felt that my grandma was right above me, telling me what to do. This feeling has always been with me even as, I must confess, I didn't always listen; and when I didn't, I paid a heavy price for it.

The pastor and I met, and I had several questions that he had trouble giving me a straight answer for. It was full of what my grandma said was "goobbidy goo." That told me things weren't on the up-and-up. So I asked him to schedule a church meeting so we could discuss what had happened and if anything could be done about it. He said he would, but everything had gone through, and everything that could have been done had been done with no positive results. I said, "That's what the struggle is all about." My grandparents wouldn't have been able to buy and own their own property unless they were up to enduring the impossible struggle. He had no answer. I took that moment to tell him that his own Bible teaches us that nothing is gained without sacrifice. He looked amazed that I said that. However, I just wanted to let him know that I wasn't one of his fools, remembering you must be respected.

I went to the office of Robert Coard on Tremont Street downtown Boston and demanded a meeting, deliberately talking loud so all could hear me saying, "Why did you agree to steal older black people's homes?" He left the office, but I knew I had left a mark because I wasn't arrested, which meant he didn't want to publicize this event. Looking back on all this, I guess this is where I started to get involved in the problems beyond myself.

The church meeting was called on a Saturday evening about 4:00 p.m. because I hadn't forgotten that most of us started having dinner around 6:30 p.m. on Saturday. It pays to know how the people live when you want to talk with them. I introduced myself, but most people knew me and said, "Welcome home. Your grandma always said you would come home." I tearfully said thank-you and began asking questions about things like how they were contacted, who represented them, how many meetings were there, how they decided what or whose home to take, and how and where they drew the line on what areas to take.

Much of these questions they couldn't answer because they were supposed to be represented by the church and the church's attorney and Robert Coard. After the meeting, some people met with me and thanked me for showing interest but said if it was God's will, let it be. I strongly felt this was something the minister told them, and later that evening, I didn't hesitate to tell him about it. Plus, I said that I felt he was in on the theft. He didn't like that much, but you know I couldn't bite my tongue. I also told him that since I believe in God, I would let Him judge him and when He's ready to send him straight to hell. Feeling somewhat satisfied, I left.

Now I must admit that there wasn't much I could do about what had already happened, but I tried to schedule a meeting with the city council, but they looked at me like I had two heads. I felt like it was a form of returning black people to slavery, like being told, "You Negros are getting to forget your place." Only because we had proved that we could do as well as white people with much less to begin with. This is the real reason why they wanted to break us up because we might become a threat for white people to deal with, and they couldn't have that.

I was seriously upset with Robert Coard, whom I was led to understand was this great supporter of the poor and a rising figure in the civil rights movement. I saw no proof of that.

Now I was beginning to understand and become much more interested in what was going on. It was like I had been swept up and landed in the middle of a fight that I thought didn't concern me. One of the important things that happened was from this moment on until the present, I found time even when I was personally going through difficult problems.

I kept trying to meet with Robert Coard, and even when he was in Roxbury for an open meeting, I would bring it up but had little or no help, mostly because people had money in their hands and couldn't see what was really going on.

I had discovered by this time that the only areas affected were Roxbury, South End, Dorchester, Mattapan—all majority black residences—not touching South Boston (where Whitey Bulger lived), Charles Town, and the north end, all areas where a majority of white residents lived. I could hear and see my grandma in her grave telling me, "Don't quit. You have much to do." I remember her saying more often than not, "You fight to win the war, not the battles." That's where patience came in, something that I still have to work on.

Finally, I want to say I did face Robert Coard in 1997 at his office while I was a student at University of Massachusetts, Boston in the College of Public and Community Services. We came to his office ironically to discuss neighborhood problems. When I arrived a few minutes later, he immediately recognized me and asked me to come sit next to him. He volunteered to say he was wrong about what had happened back in the 1960s and offered his apologies. I thought for a moment but remembered my grandma saying it is better to forgive even if you can't forget, so I took a deep breath and did.

You can tell that I didn't have a lot of respect for the man although everyone else did. I guess I saw a side of him that others didn't.

CHAPTER 13

Eminent Domain

N ow I want to take this time to give you my thoughts about how eminent domain and reparations join together. Remember our ancestors who were born around 1875 when Reconstruction was happening.

They had endured terrible times, then some, getting the so-called forty acres and a mule that the government ha and still taking back. This was our first attempt of receiving reparations. Others moved North, given a chance they could survive and do well without public assistance. They bought homes, ran businesses, and raised several children.

This was all done by our grandparents (unless you were close to seventy-five years old). Your grandparents most likely didn't fit into this group. You never got the chance to hear how things were really like. But your grandparents and mothers and fathers can remember being raised in their homes— living together, eating together, all working together, having pride in themselves and much less neighborhood crime. It's wrong that we have so much knowledge locked up in nursing homes. If we want to know the real history of our struggle, doesn't it make sense to go directly to those who have lived it? We are lazy because rather than getting off our asses and listening to them, we read about what meaningful authors tell us what it was like. That isn't much different than believing what schools teach our children. What is really needed is for us to truly return to our roots so we can be educated. Without hearing and then knowing the real truth, we can never know how to fight for justice.

Motivation comes from knowledge, and knowledge comes from listening to the truth. Doesn't this make sense? For all you young people, my best suggestion is to make time, visit the nursing homes, or listen to the elderly. Ask them what it was like to struggle for what they were able to get before the white man took it away. If you do that, then you will know what it really takes to fight for change and, more importantly, to learn how to keep it.

All this to say that eminent domain is the devil's henchman coming to steal people's homes and separate neighborhoods, especially when people begin to do well and communicate with one another rather than trying to kill one another. I can recall this maxim: when you have a happy neighborhood, you have a safe neighborhood. Take a good, hard look at what property our government chooses to take and pretend to tell us it will be better for us. Don't believe the hype.

CHAPTER 14

Reparations

N ow reparations—the government will never give us a dime if it means that they have to admit that they got their wealth from the blood of our ancestors. That would force them to admit that they used us for their pleasure. Remember what I have already pointed out what happened to the forty acres and the mule.

The only way we will be able to get this to happen is by demanding that any newly elected official make it a priority in both houses and the president to address this issue with our eyes glued to our problem because trouble doesn't take weekends off.

This is only part of the problem. Having money shouldn't make you comfortable. Remember, there is an old saying that a fool and his money will soon part. Keep this in mind: unless we have laws protecting this, we will pay a heavy price. For example, all people living in Section 8 housing will have to claim this money, then pay more rent. Health care will certainly cost more. Everything you buy will cost more, and for all those who are living in homeless shelters and on the street, where and what are they going to do with a lot of money and no affordable place to stay? Also, if you are going to use Japan as an example, keep in mind that Japan was not here when our landowner got rich. So that is why it was much easier for them to pay some form of reparations. We and, of course, the Native Americans are viewed as a natural sacrifice for the good of this great country. Never will they admit to owing us a dime. It would spell disaster. Our only chance is to vote for representatives in all three elections that we know will honor our concerns but keep a watchful eye on them to make sure that they don't take time off.

I know I have said this before, but I believe it is important enough to be repeated. Today, just look at our newly formed entrepreneurship our young black men and women are attempting to create. It reminds me of a time when we were independent. Like in the struggles of the 1960s when we boycotted and shopped at black-owned stores. It was a form of fighting that was an important assistance in our struggle for civil rights that we temporarily won. We should support this effort because it will show our children what they can do for themselves. I say I am happy to see this effort before I die.

I am only pointing out some of the problems that sudden wealth can bring without legal protections. I recall a great saying: never look a gift horse in the mouth.

Just felt I had to take this time to share my view. I have to honestly say I am looking for my reparations also. Finally let me point out that we are all children of God, made the same way. He knows that we have different opinions about things but gave us the same two eyes and ears to see and listen to others with an open mind—not to close your eyes and ears and just assume you are the only one right. All of us must sit on the bathroom stool the same way. Just a perfect example of what we all have in common. So why have we gotten so far apart as people, thinking that we are the only ones that matter and to hell with everyone else? Just remember the Lord gives, and the Lord takes away. That time may be upon us sooner than we think. Take a good look at global warming.

CHAPTER 15

Working at West Roxbury VA Hospital and My Return to My Family

N ow it was time for me to find job, mainly because my funds were running low, and I was getting restless. So for some reason, I thought I would try the West Roxbury VA Hospital. I got a job as a patient attendant assisting paraplegics in all their care. It didn't seem to bother me a bit. I asked for the night shift because it paid more money, and it gave me the day free to begin to get adjusted. Most everyone I know was beginning to smoke marijuana and drink Colt 45. It wasn't for me. I started going to the jazz clubs that were around, such as Paul's Mall and Lennie's on the Turnpike. Jazz was my favorite, but I did begin to collect a lot of the new R and B music. You look for it, I had it. Both brands.

Believe it or not, I was staying at my aunt Adelaide's house on Hutchins Street of Humboldt Avenue, right next door to Satch Sanders of the Boston Celtics, whose license plate number was Celtics 16, the way all the Celtics license plates were. As time went by, we became good friends. In time, I also became close friends with Sam Jones, plate 24, who lived on Dennison Street next door to my future wife.

I was able to go to many games for free and sit down front. In part, basketball wasn't the number one sport in Boston at that time; and when they played, there were always double headers.

Adelaide and I had settled our differences and, as time went by, became very close. I could never figure out how that happened, but I do know we never talked about the past. My two cousins Peter and Karen were there, with Karen beginning to stick to me like glue. I could never figure out why. This was the period that I spoke about earlier. To tell you the truth, she saved my bacon a few times, but she was a constant pain in the ass.

At the VA, which was a paraplegic hospital, I started at a rate of G4, which was the starting rate. But with working the night shift and overtime, I made pretty good money. I would like to point out here that I never got to be friendly with the nurses. Most of them had a private relationship with the veterans. You understand. However, wouldn't you know that trouble came along as usual?

The nursing supervisor, an elderly Irish lady, called me into her office. I thought I had done something wrong. I couldn't figure out what because I had only been there for a couple of months. She told me that she knew my father, and she hoped I would live up to his way of doing things. It was like she was sending me a message of some kind. All I knew was that I didn't like it, and as usual, I had to respond. I told her in the same tone that she had that I didn't know that my father ever worked there; however, my name was Kenneth Walden, and I only hoped to be respected that way. Then I closed the conversation by saying, "Well, would there be anything else? Because I have a patient waiting." She huffed and said, "No, that will be all." So I left, saying, "Have a great day." To tell you the truth I was very insulted by that conversation because it reminded me of when I was back in school several years earlier. I was having a flashback. I felt like things were going to be touch and go from then on, especially when she normally worked the day shift. I started thinking she was waiting for me to make a mistake or something.

As time went on, I had gotten good enough to begin specially patients in the recovery room all night, moving up to a G5 and making even more money because of the work I was doing. By the way, I always got to work at least a half hour early. This was always my way. It was especially important as, there, you always had to get an update on the patient. On one occasion, I was specializing this patient, and he died, so the doctor had pronounced him dead. I performed the morgue procedure and took him down to the morgue at 3:00 a.m. to put him in the freezer. As I was leaving, I heard a sound. At first, I thought I was hearing things, but I wasn't. I opened the freezer, and what do you think I saw? Sitting up on the gurney, looking at me, was the patient. To tell you that I was more than a bit surprised would be telling you a lie. It was like seeing that moose in Maine several years ago with almost the same results.

However, I did walk—if you believe it—over to the emergency phone and calmly asked for help. After what seemed like an hour, the doctor came and said, "Oops, I guess we made a mistake." I said what, "The hell do you mean we! Your." The doctor laughed, and we returned him to the ward, and he lasted another two weeks, then died during the day.

During the day, I started playing a lot of basketball at the old Boys Club at Dudley Street, eventually becoming a member of a neighborhood team called the Boston Stars. We were pretty good playing together as we got accustomed to one another. We started winning a lot of games. I have to say it was pretty good. They thought that I could help them a great deal because of my jumping and rebounding ability and my Bill Russell-type hook shot. When Satch came to the gym, he would begin to call me Golden Arm, a name that stuck to me for the next twenty years.

We played in several places like the yearly tournament at the Norfolk House and the usual revelry between us and Cambridge. This was always our biggest game: Roxbury against Cambridge. On the court, we hated one another. Some of the names I recall on the Cambridge team were Rosco Baker, Charlie Stead, and Ed Washington, great friends but hated opponents. It was like playing the 508 MPs in Germany. Only now I had a lot more height and experience. Also at that time, my height was considered tall. Plus, I could dunk the ball; although, in that league, you couldn't do it. But during layup drills before the game, I would always show off a bit. It was a form of intimidation. By the way, my cousin Karen would always tag along even with her homework. If I didn't take her, she would perform one of her temper tantrums. She knew it would always win. Her mother never interfered. But her brother was in his own world. He would rather stay home and, much of the time, have his friends over and play whatever.

My social life was picking up, mainly because of the absence of so many young black men who were being sent to Vietnam. I need to point out that during this time, the devil had me, or maybe I had him dancing. To my knowledge, I don't think we parted company for several years. I would call him when he was sleeping to come out to play. In fact, I believed he retired.

I made a big, big mistake. I got too involved with someone I knew before I went into the service, and she had matured into a beautiful, healthy lady. Looking back, I don't even know when I had a chance to enjoy anyone's company because my cousin was always at the door, waiting for me to go wherever I went. But it happens. To get right to the point, this lady and I got married, and I had quietly gotten an apartment on Seaver Street, or at least I thought it was a secret. She was living at the Harbor Point Apartment. When the wedding was over and we had just gotten in and hadn't even consummated our marriage, the first thing that happened was her mother coming in with this young lady and said, "Surprise! Don't forget her." I stood there with my mouth open, not being able to say a word. All of a sudden, my aunt appeared with my cousin and said, "I hope I'm interrupting. Come on, Chinky, don't you think it's time to leave?" You know what I did. Like a new recruit, I followed orders. When we got outside, my aunt laughed like hell and said, "How was your wedding? I hope you enjoyed yourself." When she finally came up for air, I asked her how she knew. She said, "Don't you think everyone talks? And your cousin knew where to find you." Karen looked at me blew me a kiss and huddled under my arm, and we went home.

By the way, my uncle Raymond was the one who brought them there. He and I had forgotten our differences also. Now that I have had a chance to look back over my relationship with my cousins, I think I know why I was looked upon and treated the way I was. It must have had something to do with the absence of their father—Peter because he knew his father and missed him, and Karen because she didn't know him, and I was the only strong male figure that she felt close to.

Soon after, my used-to-be wife filed for child support. We went to court—my aunt, my cousin, and me. She had her mother with her and the little one, of course. She was dressed like an abused woman. The judge was Eleanor Fitzpatrick, called the man-hater. I just knew I was dammed. But when my aunt said I couldn't be the father because I was in Germany at that time, in addition to me showing my discharge papers, the judge looked at my wife and said, "Young lady, that had to be the longest sexual by in history," and refused to take up the case. Of course, I breathed a sigh of relief. I don't think I saw her again, and you know my cousin had seen to it that she and my uncle had gotten all my things out of the apartment like my clothes, records, and television out before she could take them. I guess you can say that all of them together saved me from a fate worse than death, especially that little pain in the behind. As I have said, this pain was part of my life until she passed away a few years ago. God bless her.

Now back at work, I was happy to be free of my momentary lack of judgement. I Continued to do well on the job, but not to the liking of my supervisor. She would still show up on my floor at 2:00 a.m., on some occasions just to see if I was awake or doing something I shouldn't be doing. The nurses I worked with were very aware. However, fate came forward and solved the problem. She had gotten burned in a car accident, not seriously but enough that she couldn't return to work, so we all breathed a sigh of relief. Just being honest.

CHAPTER 16

Jewish Memorial Hospital

In the following year, I got a job at Jewish Memorial Hospital on Townsend Street, Roxbury, where my aunt worked on the day shift as an LPN. I was being hired to prepare patients for surgery as a nurse assistant. These things I had learned at the VA. My aunt had told me that there was an opening. A Ms. Spenser was the nurse in charge and, of course, knew my entire family. Something that I had by now gotten used to. I began to tell myself that at least it was for something good, so be happy and use it to your advantage. I was too focused on making my own name that I couldn't see the forest for a the trees. Ms. Spencer and I talked about my father a bit, and then she took me around the hospital. I was introduced to the director, Mr. Murray Fertel, a man that will become very important in my future education. It was a Jewish hospital right in the middle of Roxbury. That would never be the case today. For me, the money wasn't bad for that time.

Being a young man then, I didn't require much sleep; you know how it is. Plus, with the upbringing that I had, sleep wasn't an option—playing basketball on the weekend and all the other enjoyments, working nights at the VA, and now working days here. You would think it would be enough to stop the devil from even coming around. Wrong!

I began being well entertained at Jewish Memorial Hospital by all—and I mean all—the nurses. Or that's how I recall it.

My aunt had a field day keeping score. It got so bad it was like they needed her permission in order to see me. This wasn't easy with my cousin always interfering. She got so involved that if she didn't like you, you didn't stand a chance. She was about fourteen and couldn't be fooled. As you have already seen, she never let me forget it, especially when I told her to mind her own business. She would always answer right back that I should be glad that she made me her business because being around me was more fun than being around anyone else. She told me to my face that there wasn't a day that went by that I didn't create something new. Could you believe that, coming from a little pipsqueak?

CHAPTER 17

My Own Apartment

I finally got an apartment right across the street from the hospital. It was at 52 Townsend Street, and it was a greenhouse. I also got my first car, a brand-new 1964 Pontiac GTO. It stands out in my mind until this day. I don't know if this makes it better or worse for my social activities, but I felt like I was the new superfly.

The ladies didn't even mind being seen coming and going out of my apartment. One of the good things about it was, I always was allowed to go home to cook my meals, and I had a new dog named Duke, a Great Dane. This was only a problem because my cousin had a key to go in and take care of my dog when I wasn't able to. The problem with that sometimes was, on the weekend, she would come prepared to stay, and her mother would just laugh and say, "That's your problem."

But in my case, she was always interfering with my fun, so to be honest, I might have been approached many times. The actual action wasn't as often, perhaps for the best, now that I look back on it. I bet Karen is saying, "I saved you from having a lot of children." Or is she patting herself on the back, laughing, or both? I believe both.

In this time, I was well established at both hospitals. Plus, I was having a great time playing basketball.

My uncle, who was a thirty-third-degree Mason, was part of a group of Masons that had a black-and-white ball every fall at the Commander Hotel in Harvard Square, Cambridge. The men dressed up in tails with the black top hat, and the ladies always wore some kind of mostly white gown. I never paid any attention to this before, but now I was very much interested. The tickets were $20, but of course, I always got two for free.

My problem was, who was I going to take? One of the many problems I had was, especially when my cousin felt that it was important that she make the choice, she would say that she would be at home in bed when I got home. Now this is the brain of a child, right? She was learning too much too fast, but her mother said I was the best teacher to show her what to be aware of. Can you beat that? But I must take a moment to confess I loved that little devil, and many a time in the future, I dropped whatever I was doing to come to her aid. I guess that's what a family is all about. At last, I guess I was beginning to see this. But I still had a long way to go.

I took a young lady that would eventually become my wife: Margo Lee. She is someone that will get much more attention later in this story. I don't need to go into details telling you about how much fun it was. This young lady Margo Lee focused on me from this moment until 1969 when we got married. I will go into details later. The following year, she wasn't able to go for some reason, so I took someone else.

I decided I would take the most-talked-about young lady in the area—of course, against the wishes of my two advisors. When I picked her up, her father said, "Be sure to have her home by midnight." I of course said, "Yes, sir," and we left. At the party, we caught all the attention, as was expected. By the time I was ready to take her home in order to be on time, she started performing one of you ladies' temper tantrums. You know, I could tell what was happening. She wasn't aware that I had been well versed about this. So I picked her up and put her over my shoulders, with her kicking and screaming. With everyone there looking and laughing, I put her in my GTO and took her home. When her father came to the door, she ran in, crying like I had abused her. Her father just said, "Thank you for doing what I asked, and I knew you were going to have a rough time. I am glad that she had met her match."

I was completely surprised by it all. She tried to see me a few days later, but you know I wasn't interested. I did see her many years later. As usual, when my family found out through the grapevine, my controller Karen had a field day.

By now, I was praying for the day when some young man would catch her attention and take her off my hands, but she would always say, "Don't worry, I will never let you go because you would be in too much trouble." I often felt I was the child, and she was the teacher. I wouldn't put this curse on my worst enemy. I hope all of you men have had similar problems. We need to form a club that will discuss the abuse of being the brother of a young pain-in-the-behind sister.

I tried everything I could do, but in this case, I didn't have the answer or help from any of my family members. Now that I look back, I think it was training for the future with my own two spoiled daughters. I think the only relief I got was playing basketball because she couldn't come on the court. I know you must think most of what I am saying couldn't be true. But let me say honestly that it's all the truth; and as she got much older, things that I was called upon to help her out became even worse.

But the most important thing was, I began being the one that was always called upon to save the respect of all my family members, even today. I guess it was meant to be.

The Vietnam War was calling and drafting everyone else to fight. But I had served my time, lucky me. I was safe. I wouldn't be called back unless there was an emergency. So you see, it was like it became my obligation to keep the poor young grieving ladies happy. Or at least that's what I told myself.

However, things don't always turn out like you plan. Karen thought it was her responsibility to ride shotgun, with my Great Dane in the back seat. I was also going to the new dance studio called the Sugar Shack on Boylston Street, where my cousin couldn't go. All the named singers played there such as the O'Jays, the Staple Singers, Harold Melvin & the Blue Notes, James Brown, and Ray Charles. They came to clubs where you sat up close and personal with the singer. Our future office was at 120 Boylston Street nearby. It allowed me to get acquainted with the heartbroken ladies— and don't forget the jazz clubs.

But I believed my grandma sent Karen to prevent the devil from taking me straight to hell. This may sound crazy, but if you have a better answer, then use it.

CHAPTER 18

Becoming Actively Involved in the Civil Rights Movement

Karen was still on my heels. It wasn't all fun and games. I took the matter of our civil rights problem seriously. I even became cochair of the Roxbury District of the Urban League, along with Chuck Turner. We had an office on Washington Street, near Elligiston Station. At this time, racial tensions began, beginning with the busing problem. I was driving to Howard University for a meeting, something I enjoyed because it gave me a chance to open up my car on the turnpike. At that time, there wasn't the same road restrictions. Gas was $0.25 cents per gallon, and car insurance was around $400. I could make it to DC in a little more than five and a half hours—believe me, sometimes without my shotgun rider.

You are probably wondering when she had time for school or even doing whatever young ladies did. Well, the only answer I have is, she did most of her homework at my house. Mostly on weekends and holidays.

To some degree, you would think that God was punishing me for some great wrong. But one thing I was still learning—family was family, and that's it.

When I was going to DC and participated in civil rights discussions, I often took Karen along, in part to educate her about what was happening and to just stop her from getting on my last good nerve. I became friends with a beautiful young lady who, of course, my cousin could recognize that I was more than a little interested. She was a student active in civil rights affairs at Howard University. Over time, we got to know each other pretty good, but not sexually. By the fall, we met and planned to hook up the following weekend at a big dance that they were having.

Remember that there was a big shortage of men, and I was driving the "I got you" car. So the following week, I had asked for permission from my caretaker to go alone. She gave me permission with caveats. I had to agree to take her and her mom to the drive-in movie when I returned, or I would see one of her temper tantrums. So I promised anything just to get away.

I arrived and picked this young lady up. We went to my aunt's house, where I could take a shower. As I was in the bathroom, I could hear them laughing, so I just knew things were going to be great. Not so fast. When I came into the kitchen, they couldn't stop laughing; but after what seemed like a while, my aunt told me that she was my second cousin from Poughkeepsie, New York. I couldn't believe it. My aunt said, "Well, at least you have good taste." Well, you know how that dance went. She held me close to her side just like I was home with Karen, introducing me to everyone, not letting me dance with no one; and when the dance was over, she had the nerve to walk me to my car, give me a big hug, and a kiss on the cheek so everyone could see, and waved me goodbye. Now you know, when I got home, everyone knew already. My cousin still made me keep my promise.

CHAPTER 19

Boston School Segregation

In Boston, there was a serious school-busing problem intended to ease the attendance of all poor black students being in the same class, getting a substance education. The major trouble came from the school committee's attempt to bus children into South Boston, a section of Boston that was 99 percent white for more than one hundred years. Plus, a lady, Louise Day Hicks, a member of the board, was 100 percent against it. So there were riots and fights. I must say that there was the question of whether she was a racist or just didn't want more children coming to South Boston to schools that were already crowded. Plus, they were short of good teachers also. Our chant was, "Bus better teachers, not our children!" No one wanted our children to leave our safe community; this was like what was happening in Alabama with the black college students. Our chant, as usual, got little to no media attention. The law was passed in 1965 (Appendix 4 will explain the details). Government involvement came in the early 1970s

It was personal for me because it affected my cousins Peter and Karen. I became even more involved in the civil rights movement. As a result of this and being a member of the Urban League, I was able to voice my opinion at the city council meetings, along with members of the NAACP. Come to think about it, I don't recall seeing Robert Coard at any of our meetings. Although, I must admit, I could be biased. However, I am trying to point out that we did have what you call many fence dwellers. The national call became, "Bus Better Teachers, not Our Children."

The METCO program where I was also an executive board member was started. This organization was able to get federal funds to have children bussed to suburban schools. It didn't happen overnight. I should point out that this is different than the picture that Kamala Harris portrayed. I believe she benefited by busing (Appendix 6, "History of Program" will explain details). This was just one more problem that incited me to the cause of racial injustice. This problem has continued.

In the early 1970s, even with the federal government involved, things haven't changed much. You will find it very difficult to find anything written by the media concerning Boston. Why? I just have my suspicions. It was because it was Boston, and here, it was supposed to be a state that cared about the welfare of the "poor black people." Sounds familiar. Also, I don't recall Robert Coard yelling out for black justice.

CHAPTER 20

My Thoughts on Racism

I received my education about racism from one of the best teachers on the subject—my grandma. She explained her view to me many times, pointing out many examples. Her view was that all forms of racism are based on profit, and no group has been affected worse than the Native Americans. They were here first and never invited the white man. They were living well off the land that was free and gave all who lived there enough to provide for their needs—the only group that has been treated worse than us, the so-called niggers. It is true that we were forced to be here and must be compensated. However, racism in its purest form has been applied to even white people against white people from the time of the pilgrims. Every ethnic group that has come to these shores have been set against by the group of people already here. The landowners found it much more profitable to band together and steal niggers from their homeland and use us to provide their wealth. This happened because they found that trying to use Native Americans wasn't working out. Remember, they are called native-born because they are the only ones that can truly claim that they belong here when the first white man came.

However, as time went on and so many other white people were coming to America, those who had already been here denied anyone else that arrived an introduction. For example, New York, when immigrants from Europe were coming here in droves This is a good example of a reason that whites who are really in the same predicament as us are encouraged to attack us in order to protect what they really don't have themselves, marching around and saying such things as, "Jews will not replace us," never thinking how many American-born Jews are getting much help from the government. Just because the media doesn't cover it, it doesn't mean that it isn't happening. Remember the Holocaust. Can we honestly say they haven't suffered?

They do not realize that it is the very people that they are protecting. The rich are responsible for the poverty of all of us. The racial chants are in part because they fear that we will get reparations for what their relatives have done and they will have to come face to face with the truth. Remember, as my grandma showed me, when the government gives you something, there is always a profitable reason for them to give it to you. For example, when the government granted us forty acres in the late 1860s and what was popular to say a mule, it wasn't because they want to help us. Hell no. It was first to appear like they wanted to help us, but in truth, when we started to make it work out for us, the government used their laws to take it away and still does. It's like we clear the land and plant the seeds, and here they come in time to reap what we have sown. Just ask our black farm owners a subject that gets little to no media coverage during the Trump administration. Working sometimes by using imminent domain or anything else they feel to stop us from reaching a point of being successful.

Nothing is free. You must always guard against the devil demanding payment. That's what she always said. She should know. She has been there and had been affected by it. By the way, how many of us remember Trump saying that he would stop producing his clothing in China and Mexico? Well, ask yourself, has he stopped? This is just a small example of him telling us one thing and doing whatever makes him a profit. I point this out in order to hopefully let our white brothers and sisters see what I am talking about when I say he is just using you.

Racism must be fought the way the wind and serf treat rocks and mountains, being pounded regularly, until eventually causing it to be reshaped. Remember what I said earlier. We are all God's children and will be called to answer for what we do with the life and gifts that he has given us.

For example, take today's issue of women's control over their own bodies. Is there any woman involved in the decision? Hell no. What if we men were told how we can govern our penis? Would we like that? Hell no. Now the real question is, isn't that a form of racism made by one group of people over the concerns of others? No color involved. That's the real facts. Now take time and think about that.

When you look at it the way I have shown it, you can see that most of us throughout America are victims of racism. So we all should stop the bullshit and stop pointing the finger of blame on one another and focus our anger on who makes the decision and the law that has us suffering.

I want to take a moment to let you know my thinking about why our white radical brothers and sisters are acting the way they are. They don't focus on the real problems like how they serve this country like everyone else and not get the proper health care that they deserve, being promised by past and present government administrations that they are held highly in our government's treatment of all who served, getting them to think that others are getting their benefits. Not true. They should stop and think for themselves for a minute and see the forest, not just the trees. This is just one of the many examples that our government blindfolds us into thinking that others are infringing on their rights when, in truth, they are played into doing the government's business. Trump didn't start this problem, but he has improved it. I beg all of us who are being denied our basic rights to stop fighting one another and focus our attention on who is really to blame. When or if they take the time to think about their concerns, they will find that they are focusing their anger on the wrong people.

When we finally won a battle for civil rights, in truth the war was far from over. If you need proof, just look at us now. We are losing our voting rights that we thought we had. Ask yourself, why is that? I will tell you what I think. It's because we went home feeling great, forgetting that it was only a battle, not the war. The government knew if they removed MLK from the movement, we would retreat. I blame us for that. But we can learn from it. One lesson is that winning a battle doesn't mean winning the war.

The government was just regrouping. Now since we have gotten drunk off our past small victory, they have returned in force and are taking it away, along with any hopes of getting more. We always want to retire after a small victory rather than getting prepared for it to be taken away. Will we ever learn?

Today we need to advance the efforts of all our young black entrepreneurs that are taking it upon themselves to be their own boss—just like our ancestors took it upon themselves to grow their own food, have their own grocery store, barbershop, hair salon. Now open your eyes and see all of our people that are trying to advance us again to when we showed our independence. Remember the 1960s when we dressed in dashiki and Afros. We must show our children that they can have the same thing that white people have (their own place of employment) and have self-respect.

Once again, we have a rare opportunity to get back in the fight. Ask yourself, are you prepared to do more than what we have done in the past? If not, then we should quit right now, just give up. If we are prepared to fight, we must create different methods. Like being sure that we pick the right candidate in all three elections. Get out and vote, and most importantly, be sure that our elected officials focus on what we voted them in for each and every day. Remember once again that your problems don't take a day off. For those of us that have lived from the 1940s until today, remember what I am talking about when I say that you don't gain a damn thing worthwhile without a fight, and you don't keep it without a bigger fight. I hope I've touched a nerve. All power to the people.

CHAPTER 21

Beginning of Our Union

The director of Jewish Memorial Hospital was Murry Fertel. This man became very important in my life for sometime. After working at the hospital for more than a year, early in 1966, I became much more interested in the way our people were being treated in the community and, in particular, in the workforce. There was a small group of us. Zema Thorton was the most important member of our union. She was the rock that held me together. I remember her being the spitting image of my grandma in every way. She helped me decide on every important issue and worked at this hospital until 1980. Judy Cooper worked at Jewish Memorial Hospital and became vice president and district executive board member until 1980. Diane Dean worked at university hospital and a district executive board member who was the free spirit that brought the energy of our future health-care workers abroad, and Ms. Witmore worked at a Springfield nursing home and a district executive board member who, without her knowledge of the real concerns of the nursing home's problems, we would not be here. All these strong-willed ladies never collected a dime from our union. They kept their jobs where they worked and volunteered their time. Finally, of course, there were my two coworkers whose methods I didn't agree with, but I must say that they did bring members to the table. It was just that I had to teach them what our union was all about. All these women were very instrumental in our efforts to have a union.

CHAPTER 22

Negotiations—the National Union of Hospital and Health Care

Our conversations centered around how and when we were paid, vacation, and of course, sick time It wasn't long when the subject came up about organizing. I don't remember how or who first mentioned it. We gradually got others to get involved, and before we knew it, we had formed what later became our first organized nonprofessional health-care union. I must point out our issues had nothing to do with the racial problems around us. Mr. Fertel treated all of us fair in that regard. Or at least we felt that way. Plus, keep in mind we were located in the heart of Roxbury.

I had some knowledge of how and what a union was from my days with my truck-driving uncle. However, I knew we weren't going to be fighting and busting heads. We somehow got the National Labor Relations Board to conduct an election. As I recall, there wasn't even a law at that time giving us the right to vote. However, we won the right to have union representation. So now it began.

This was one of the many difficult moments. None of us knew a damn thing about negotiating a union contract, but it was apparent that the two main members (who thought all you had to do was make demands) thought they knew and constantly disagreed with me and didn't want me involved. However, maybe because I had shown more calm and common sense, the other members voted me to lead the negotiations. I had to quickly gather all the information I could. This meant going to the teamsters' office and introducing myself by using my past affiliation with them back in the 1950s. This went over very well because some of the now officers of this union just happened to be there during some of those struggles. Because of that, they gave me the basic information I needed and helped me behind the scenes throughout this millstone negotiation.

The negotiation committee was made up of my two rivals, Zelma Thorton, Judy Cooper, and a few others that I can't remember. Mr. Fertel didn't come off like he was totally against us. In fact, he was very cooperative, considering this was 1966, a date that should always be remembered, partly because it was the height of the civil rights movement and the beginning of our own racial busing problems. The only positive thing at that particular time was, all of us were black. We hadn't even thought about growing to include others. It was like the blind leading the blind.

Our first union contract in 1966 included pay every week instead of every two weeks, sick leave, and some vacation time and, most importantly, a raise and paid overtime after forty hours a week. I thought it was equally important to negotiate a management rights clause. I was told that without a strong employee protection clause, you will not last or have a guaranteed contract, especially as this was our first. It wasn't as important to the others because they were more focused on money.

But over time, as I became the negotiator for all contracts, it became very important. As at least as important as money. I was appointed chief shop steward of our contract. This also didn't go well with my so-called friends. This attitude continued throughout my tenure.

I do recall it caused a long-lasting effect in our future efforts. Some of the particulars of our first contact was, any benefit currently used by the members will still stand wages went to $5.00 an hour for new employees a week and $5.50 per hour or a $0.50 cent raise for all present employees, whichever was higher. Sick leave was two weeks a year after one year of working, three weeks after five years of working. With the sick leave, you could get one day for every month after your first year. If you were a current employee, you got a week or whatever was in the contract. If you wished, you could join the hospital pension plan. I had already joined. Union dues were $3 per month. Plus, all employees started at this point with a clean work record overtime after working forty hours when you had to work forty-four hours before and getting paid every two weeks—now every week—and a union bulletin board.

CHAPTER 23

Getting Acquainted with Marrey Fertel

About this time, the director Mr. Fertel called me into his office and talked to me in a very friendly voice. That caught my attention because I didn't expect that kind of exchange that made me feel we were on equal footing. He said, in general, that he would like me to drive him to meetings and be available to pick up and deliver things related to the hospital because he was getting complaints from the staff that wasn't in our union. I quickly agreed, mostly because it would give me a lot of free time for myself with the car to drive. Lucky again. I was responsible for picking up medical, maintenance, delivery of pay information to the check-making company and pick it up every Thursday. I took all banking transactions and signed for them at the Shawmut Bank in Dudley Street every week.

I drove Mr. Fertel and a man named Joe Lensey, who was the owner of Cutty Sark Liquors on Commonwealth Avenue. I took them to the hotel in Park Street at 11:30 a.m. every Wednesday and picked them up at 1:00 p.m. and drove them back. This was my job at the hospital for the next few years. I learned much more than I could be taught at school during this time, especially driving Mr. Fertel. We became very close. I always kept my mouth shut and listened while they were discussing business. You would be surprised how much you can learn by just staying quiet. I later learned that Mr. Lensey was once a part of the Mafia. I found it hard to believe, but it wasn't my concern.

On occasion, I had to deliver material to Mr. Lensey at his office and told to give it to him. I had enough sense not to look at it and only put it in his hands. Once I went to his office, and his secretary said she would take it for him. I quickly responded, "No disrespect, but I was told to put it in his hands, so I will wait." She smiled and called him out of his meeting, telling him what I had said. He laughed and complimented me and said I listened well and thanked me. After that, he would call and have me take things to his wife, who lived in Wellesley, Massachusetts.

Mr. Fertel and I had many personal conversations about a person's family history and the importance of it, as well as the importance of knowing all you can about a subject before you enter a conversation. We talked about much more, but those two things stand out. He even allowed me to attend classes at the University of Massachusetts, Boston, at Park Square called College of Public and Community Service to learn more about negotiations. This was very unusual for a man that I was sitting across the table negotiating contracts. He would say, "If you are going to do this, then be the best and learn all the parts of negotiating. Both sides." I found all of this very important as time went by. We never got into an exchange of what was going on in the hospital. It was a very strange relationship. In the winter, I had asked if it was okay to arrange to pick up the dietary workers at Elgaliston Station between 6:00 and 6:30 a.m., especially in the wintertime.

The hospital sat on a steep hill, and it was difficult. It meant that during that time, I would be available from 6:00 a.m., going up and down the hill that time of the year. So it was every morning that I picked up people. I really didn't mind because it was teaching me about people in general, which I found very useful as I moved on with negotiating union contracts.

CHAPTER 24

Internal Problems

I began having problems with my two closest organizers as soon as I started my job driving. They started talking about me being too close to management and that I was selling them out. This opposition happened the entire time I was with the union. None of our fellow workers thought anything about it and quickly told them to shut the fuck up. I wasn't aware of it at first but soon found out that it was Zelma Thorton and Judy Cooper who settled it.

I constantly had to settle the problems this created. They would try to make the workers feel that with a contract, they didn't have to listen to the bosses. That was the biggest difference between me and them that caused serious differences. I tried to show them that a contract was meant to improve working conditions for all, including patient care, that without the care of the patients, there would be no jobs. They were good at getting people interested in joining our union but no interest in educating them about the complete benefits outside of more money and less abuse. I had to handle a grievance a day in the first year, sometimes having to tell workers when they were wrong. Not an easy job, but I found it necessary if we were ever going to convince people of what we really stood for—something that was hard for some of us to understand.

I was well aware of the consequences of what our future would be trying to organize white and black nonprofessional health-care employee during the serious racial divides that was going on here in Boston. Remember, we had the serious school busing problem, where the Boston schools committee decided to bus black students to South Boston white schools, in the same area that Whitey Bulger was operating. That was just some of my problems with my own coworkers. I had a difficult time to think beyond the nose on their face. Besides, I was told by Elliot Small later in our organizing efforts that one of them couldn't read well. If it was true, it was covered up very well. But many black people still couldn't read that well. In the appendix that includes my picture and my interview with the Boston Globe, you will find a report submitted by Elliot Small to the National Union explaining the internal problems regarding our organizing and my being in charge of Boston hospitals where my relationship with the others was in conflict.

CHAPTER 25

Family Ties

Maybe because of what was happening here at home, or maybe I just wanted it. But I felt it was time for me to try reuniting with all of my Family. If I wanted to bet, I would bet on the fact that my grandma encouraged it. I was starting to feel that I was now the head of at least the Boston family, mostly because of decisions that I was beginning to be asked about from both my aunt and uncle. Plus, you can never forget Karen. However, there was some good news. Karen was going to school in Lexington by way of the METCO program, which meant that she was beginning to develop friendships. Boy, was I happy. God had answered my prayers. But it didn't mean that I was completely free. That was too good to hope for.

I decided to go to Detroit, where both my mother, father, and his wife and children were living. It is important to note what you feel about your family. You come to realize that they are family, and if you have any feelings and not afraid to face it, you will feel greatly relieved. I went alone this time. Believe me or not, I first visited my father and his family. I had a pretty good time. Didn't stay long. I guess it was because it was so strange. However, I came away feeling like a mountain had been lifted off my shoulders.

Then I went to visit my mom. I had known where she lived for some time because she was always in contact with my cousin Joan. I never asked why. I got there, and as usual, it wasn't long before I had to open my mouth.

A man was there visiting. I guess he and my mom were in the kitchen, drinking and smoking. That didn't bother me, but what he said did. He told, not asked, my sister (who was around fifteen years old) to come over and give him a kiss and sit on his knee. Well, you know what happened next. First, I threw him out, and it wasn't even my house. I told my mom what I thought and then told my sister to pack her bags because she was coming back to Boston with me. Thank God it was the beginning of the summer school break. I didn't even give her time to come up with a response because anyone who knows me knows I don't waste time with small talk.

We left, and I told my mom that I would call her later when I had to calm down. We drove by way of the Canadian 401, which is the no-speed-limit way to get to Buffalo, Niagara Falls. She wanted to see it. By the way, it was about two hours into our trip when she finally got up enough nerve to speak. After all, she hadn't seen me since she was born.

We talked all the way home, and she became comfortable and eventually told me that I was God-sent. I told her, "I am here now and will always be there to protect you." I also told her that she didn't have to explain anything to me. After all, I knew what it was like.

On the way, it finally dawned on me that I had to figure out what to do with her. After all, I wasn't exactly prepared for all this. But believe me or not, I felt that with my grandma's help, there would be a way. We arrived, and I fixed something for us to eat because I knew she didn't know how to cook anything. Somehow I could see what her life was like. She was tall for her age like myself, but the way she talked, I could see that she was totally dependent upon our mother. At her age, she showed no teenage attitude. So all of this told me that my work was cut out for me. I gave her my bedroom. At first, she was nervous and fell asleep in my arms like a little girl. After a couple of days, it was time for me to go back to work.

So I decided to try and get her a summer job at the hospital being a pinky. That's what the young teenage girls were called who fed and assisted the patients. The hospital hired the neighborhood young ladies in the summertime, something that doesn't happen today. Mr. Fertel even employed black young Ladies as far away as Tuskegee, Alabama.

I had no problem getting her the job; plus, Karen worked there also. Because Karen was there, it gave me a big break. So you see, there was an answer.

After a few weeks, she began to change. She was more outspoken, something I think she picked up from Karen, much more than what I bargained for. They even started hanging out together. However, you know I was nervous, like a father caring for his daughter. Karen knew this and teased me about it all the time. By the end of the first month, I began to feel more comfortable about everything. I even called my mom and was able to talk to her calmly. I told her that everything was fine and even let my sister talk to her. We stopped talking on what you may call a mutual understanding.

Now the most important thing I remember that happened the next month was, she became a young lady at home with me, and I knew nothing about what to do. I called across the street and got my aunt on the phone, talking as if my house was on fire. She laughed and came over and took charge, asking me if I needed some smelling salts. I don't know how any man can deal with that without some education. I said, "Never again." Of course, it couldn't be kept secret. Everyone in the hospital found out about it. You know I was embarrassed.

The rest of the summer went well. She was able to learn how to go shopping and many other things young ladies do. That included getting on my nerves every now and then. I guess you have to take the good and the bad, something I was still learning. I had to look at myself and say, "I am no prize." This turned out to be a very important personal time in my life. I had come to learn much about personal pain, anger, and the true power of forgiveness. I was very happy that I was available to help my baby sister when she needed it most. I truly felt my grandma's hand rubbing the top of my head. That was all I needed.

By late August, it was time to take her home. So off the three of us went. You didn't think that my pain-in-the-butt cousin wasn't riding with us. Her reasoning was that she didn't want me to fall asleep on the road. The truth was, quietly, I was teaching her how to drive. Sound familiar?

When we arrived, she showed that she had grown up a lot, even wearing her hair differently. She had about $500, and I made sure that it remained hers. I made it clear what my thinking was and made my sister promise me that if she needed me, she would call, no matter what anyone said.

We didn't stay because it was early Sunday morning, about 1:00 a.m., and I had to be at work Monday morning. During those days, who needed sleep? Because it was an open highway most of the way, I let Karen drive. She could handle my GTO very well, but I refused to tell her because I knew what that would lead to. Remember, I've been down that road. I really felt like a different person by the time we got back to Boston. It wasn't long after she was getting involved with teenage activity in part of her attending Lexington High School. Thank God for His wisdom. That same year, I won $5,000 playing the street lottery and purchased a T-Bird convertible, the same one that is in the movie Thelma & Louise.

CHAPTER 26

The Beginning of Using Our Contract

I want to point out that my biggest problem with our joining the National Union of Hospital and Health Care
Employee Union 1199 was their primary interest concerning
our union dues and not the more important interests of our
members. You will see me call attention to our union-dues
problem throughout this period.

Remember, we were new; and even though we were
paying some dues for our protection and the hope that we
would grow, our members hadn't reached an economic point
where a high amount of union dues could be obtained to
support our own progress, let alone supporting a union in
general, especially when we would have no say in how our
union money would be spent.

In late 1967, my cofounders had, without telling us,
joined us with the National Union of Hospital and Health
Care Employee Union 1199 from New York—without discussing this with us as the affected members. They came
into our hospital and asked the members to vote for an
increase in our union dues. Remember, we were brand-
new; we were learning on the fly. Our dues were kept in
the Dudley Street Bank where it took two signatures to

get any money out. We didn't count on anything like this happening.

At first, I was out of the building running errands. When I came back, Zelma Thorton was there confronting them, and when I found out, I took the ballot box and burned it in the hospital's furnace. I told them that there would be no due increase without a meeting with our members first. This was just the first of the most serious disagreements I had during the whole time of my involvement in this marriage with this union.

Around that same time later, I was invited to a meeting in New York with the president of the National Union of Hospitals and Health Care Employee District 1199: President Leon Davis and Henry Nicholis.

We didn't get off on a good foot from the beginning, but as usual, the others took it all in without question. I had serious questions about the whole thing in general. First, joining with them, and the benefits we received was too good to be without consequences. My grandma and Mr. Fertel taught me that. However, let me say, at this time, I was outvoted, so the die was cast. They opened up the office at 120 Boylston Street at downtown Boston with a complete organizing staff of New York organizers, along with my two cofounders. A man named Elliot Smalls was in charge. I mentioned him because of the importance he played in our union as we progressed. I stayed at the hospital but was called on to negotiate the new contracts that were gained through their organizing efforts. Being paid by the national union for my efforts was something that I had demanded, thinking that it would allow me to have some involvement in what was going on.

We could only unionize the nursing home because the state wasn't paying much attention to this. There wasn't even a law legally giving us this right. Because of this, there were many arrests of our organizers, which meant paying for their efforts.

These were the hardest contracts that I negotiated because our union organizers didn't understand or check out whether or not the nursing home's money was dependent upon what the rate-setting commission gave them to operate. Something I found out before starting negotiating that played an important part of how you organize workers. For example, it's wrong to set workers to think that wages are the owners' fault. It doesn't help the spirit of improving working conditions; workers most still care about the welfare of the patients under their care. All these things and more give the workers a clearer understanding of what the union stands for. This was one of many serious problems I felt would happen and did because the new affiliate was more interested in getting numbers for dues than anything else.

I am not against paying union dues. It is like the gas that helps the engine run the car. However, I know that you must always keep your eye on who is driving because if you don't, you may be headed in the wrong direction. Union dues can be used to advanced either the members' concerns or for those who handle the dues. That is why I had always stayed focused on who was handling the dues because most trouble occurs when there is a money problem.

The operation of the nursing home was dependent upon what the rate-setting commission granted the nursing home, and at that time, it mostly depended where the nursing home was—I mean if it was located in a black or white area. Remember, this was the time of the civil rights movement. Boston wasn't isolated. Plus, the New York Organizers were all white, something that I had mentioned when I met with Leon Davis, but he chose not to listen. I do want to point out that I always felt a bit of racism because they just didn't pay attention to the real needs of black employees, just what it meant for their interest. I may be wrong, but it was my view.

CHAPTER 27

Problems with New York—the Beginning

During that same year, things seemed to get worse. You would think that since we all basically wanted to make our union grow, we would be able to solve our differences. But it got worse. They didn't want me involved, and I didn't want them involved either. It was amazing that we increased our members at all. I had to stomach the insults, and I guess they had to put up with my presence. Elliot Smalls seemed not able to solve the problems or didn't care or knew how.

The National Union had sent a man named Bob Mulinkamp, whom I first met in New York; he came to Boston quite regularly, acting like he was in charge. I asked Elliot Smalls what was his responsibility but didn't get a real answer.

I continued to negotiate contracts (Appendix 7 will show my name saying that I was responsible for negotiations of contracts). I had to meet with the workers before meeting with management about what union stood for and having a small group of the various departments under our union to join me in the negotiating, hopefully getting them to understand what it was all about to negotiate a contract. It seemed to have worked to some degree because having the effective members involved with their own contract helps their members see the whole picture. It was a great way for them to understand that the boss is not entirely at fault. This the truth. If anyone puts another idea in your head, the first thing to do is run the other way, especially in health-care contracts, because they are not there to protect your interest. I say this not because I was opposed to our affiliation with the national union alone, but Unions in general. It's because if a union is to serve and grow, it should always negotiate, not dictate, the terms. Because in our union, the bottom line was patient care. That makes organizing much different. Besides, even though I was in charge of the negotiation, the contract wasn't official until the National Union signed it. To me, it meant that they took all the credit.

If you are not careful, you could cause the unforeseen closing of the very place your workers are employed.

This is one of the many things I kept in mind while negotiating union contracts. As time went on, my ideas working caused the organizers, to some degree, to admit that my way was right, but not enough to change their ways, mostly because I was the one doing it. I then started to really see that if I truly felt I was right, I had to stand my ground and be the man that I felt I was and act like it. This brought me to think, I am who I am. If you believe what you are doing is honest and correct, then never yield. Many times I found myself at odds with almost everyone around me, but as time went on, I got more independent—mostly because Zelma Thorton was always close by to counsel me. That never meant that I didn't listen to others, but the final decision was mine.

I went with what I believe was the best decision for the members and the health-care facility because you can't have one without the other. Fortunately, I seem to have made, for the most part, the correct decisions. Remember, we were only to organize black-owned nursing homes, and there was still a lot of racial tension.

CHAPTER 28

Negotiation Skills

I want to tell you what I developed as my style of negotiating a health-care contract. It is much different than negotiating any other kind of contract because you are dealing with patients' care. Not things. No matter how bad working conditions are for the workers, it is as equally important to consider the welfare of the patients Because without them, you have nothing to negotiate. The first things you need to do is meet with the head of the facility and talk as much as you can about the benefits the facility would receive by having a union, especially if it was the first contract, always with someone from the facility, and point out what improvements were made at another facility that had our union, which was important to ease the tensions. Then ask for materials such as employees' payroll records, company policy regarding employment, any rules or regulations that employees were asked to sign or were made aware of when they were hired. If they had any present benefits, such as sick leave, vacations, or any others, along with what was their patients' payment rate from the rate-setting commission, offer assistance in discussing their problem. These are all important to be reviewed before being submitted as the union's proposal. If you don't know these things, then you are negotiating in the blind. I had to do this by myself because the others didn't know how or weren't interested. Even Elliot Smalls, who—if you remember—was sent here to represent the National Union, wasn't good at it. This strengthens my view that New York didn't have our interest at heart.

That was the main reason why I was needed, even though I wasn't wanted. After going through the material with the help of some of the effective employees, I would submit our proposal, always beginning with the management rights clause—the most important part of negotiation. Without members' protection, no contract is worth its salt. We made sure that we got "just cause" clause. Management could suspend employees without just cause, doing away with the action of using arbitrary, capricious, or bad faith. This was routine in every contract: the words that stated that at the hiring date of that particular worker, anything that was in their file that could be used against that member would be erased. Also, if any facility was to expand, the workers will be automatically be a part of the bargaining unit; and if it was to close, it must give a three months' notice and give their employees a month's severance pay. Also if the facility were to be sold, the new owner must honor the union's contract. Last, but by no means least, a union bulletin board. This entire article was shown to be the most important building block for all contracts. If you couldn't show this to be the most important article, then nothing else was guaranteed.

The other articles dealt with wages, benefits, and the conditions of employment, also, wherever possible, management assistance in helping employees that wish to advance in the field of health care, which would be a benefit to the facility. As time went on, this became very useful to especially hospitals—one more thing that hadn't even been tried elsewhere. Workers would always say, "This is harder than I thought," which told me that I was getting my point across. I always tried to advance our popularity, which I thought was a great way to encourage more people to join our union. After all, we were new and untested. Plus, sooner or later, we would have to include white employees, and that would be very difficult with all the racism that was around. I made sure that they kept their coworkers informed. I always had a member a shop steward come with me when I met with management.

This way I was always protected from any rumors about my dealing with management without their knowledge—very important, especially during this time. Remember, the civil rights movement was still going on with a lot of different feelings running high even in Massachusetts. My important view and care about patients' care was always a selling point for any contacts. As time went by, I began teaching the shop stewards how to administer their contract using the grievance procedure, making sure that the member who they were filling the grievance for always did what was asked first and then file a grievance. In this way, management couldn't say that we had endangered the patient. Plus, it gave the members much more security. Also, it allowed me the time I needed to do other things without always being on the phone with the shop steward, trying to convince them that they could handle it.

CHAPTER 29

Murry Fertel's Advice

I want to point out who played a very important role in my knowledge and ability to do this job so well. It was no other than Mr. Fertel. This was the meaning of our relationship. He'd always say, "If you are going to do a job, be the best at it, and always be truthful and trustworthy." No one knew this, and it wouldn't have helped if they did. I negotiated many contracts with Mr. Fertel up till 1980.

My last negotiating gave the workers a contract that was a milestone in our union: giving workers a salary of at least $ 12 an hour; plus, in the second year, it would move up to $15. There was also vacation up to a month for those who had been there for at least ten years—sick leave and vacation time that could be sold back if needed, and willing to promote and help educate people in the Union for jobs that would require them to leave the union.

This contract that started implementing these benefits back in 1973 were used for all contacts thereafter. The biggest benefits for the hospital were no turnover and the help of the cooperation of the union made it easier to get a yearly higher rate from the rate-setting commission. There were many more benefits received from our relationship as time went by. I could never completely understand why he treated me the way he did, but one day, he said he saw something special in me. It reminded me of what my grandma had said. Now late in my life, looking back, I can see the meaning of what they meant. A lesson for all of us: listen to your grandparents; they always know best.

CHAPTER 30

Big Changes

D uring this time, Elliot Smalls came to the hospital and talked to Mr. Fertel, asking him to allow me to leave the hospital and join the staff of the Union. Mr. Fertel called me into his office and asked Elliot to step outside. He told me what Elliot wanted and asked me what I thought and to think about it before making a decision. "Tell him that you need time to consider it. That would make him know that they need you more than you need them and would help make your relationship with others much easier." It made sense to me, so I was smart enough to take his advice and followed it. In about two weeks, I had made my decision to go on the union staff.

I made it clear that I wanted no interference with the staff. He told me that he wanted me to be the secretary-treasurer of the union, doing all contract negotiations and appointing me to be the Massachusetts representative of the National Health Care Employee Union 1199 in New York as a vice president on the AFL CEO executive board. Because we were an area member of the National Union, we didn't have district elections yet. The one good thing that happened was, the National Labor Relations Board had just ruled in favor of allowing us to unionize nonprofessional health-care employees (Appendix 8 will show a statement I gave the Boston Globe in 1973).

This meant, however, that we had no say in what or how our union dues were being spent because all dues went to the Nation Union office before being given to any area. Plus, all our contracts had to be finalized by the National Union. One more reason I was uncomfortable about being under their rule. But this meant that I would meet with the National Union's president monthly with equal say on all things. I also started asking to check the books. You knew that was like letting the Fox into the Hen House and closing the door behind you.

At my first meeting, I brought up my thoughts about if the National Union had a vision of what they hoped to gain. There wasn't an answer. So I continued my thoughts, asking if they had ever tried to bring members together at a big event, not just us officers. Because of the regular dues-paying member, we are looked upon like management. So if we don't keep them involved and happy, then we may as well close up shop. Also had the National Union thought of bringing out of the rank-and-file members to organize because no one understands what a worker goes through better, and finally negotiating contracts that allowed members to be able to bid for a better job even if it meant losing that member. These are goals that would encourage a worker to join. I know because I just came out of the rank and file. All these things were strange to our executive board members. Of course, they not taken up.

Because we were headquartered in New York during a time of criminal influence, me coming from Boston was very suspicious. As time goes on, you hopefully see why.

I noticed immediately that I was the only member who directly worked at a hospital. This raised the question of how much they knew or was concerned for the members they represented. Also, once again I noticed Bob Mulinkamp at the meetings with no official reason for being there. This always bothered me. Call it what you want, but it doesn't change my view.

CHAPTER 31

National Arbitration Advisory Board

Almost immediately after coming on board, I got myself appointed to the Massachusetts American Arbitration Advisory Board located at 495 Washington Street, Downtown Boston. I served on that Commission until 1980. Handling contracts like ours was new to the Arbitration Board. My letter from my attorney that is explained in my pension request will provide an explanation for my involvement.

It was very important that I show that we were as interested in our patients' care as we were about our benefits, introducing the fact that in arbitration, it didn't take a lawyer to represent the union. All that is required is a union member understanding the grievance and the remedy being asked for, mainly because there was nothing in the law at that time that any lawyer could refer to.

This became a cost-saving device and added pride to the union members that was effective.

I started holding classes for the instructions of shop stewards, teaching them how to go about handling grievances. The first thing was to make sure that what the workers were asked to do was done so that management couldn't say that patient care was affected. Then start the grievance immediately using the proper article order and time of evens, making it clear that it was in violation of the management rights clause. Most grievances were taken care of through that level of communication. Very rarely as time went on did we have to go to arbitration. It was used most often in our first contact because the employer and the union members were new to this new form of working.

For example, Jamaica Towers Nursing Home in Dorchester, Massachusetts. When we got our first contract, the owner did his best not to honor it. Even after the union was able to get an increase in his rate, he refused to pay the members the proper wages. We had to follow our grievance procedure, which meant taking each step from the shop steward filing first, then the office filing for arbitration. He tried to avoid the procedure by not accepting the grievance, so I had to send it to him by register letter. When the arbitration began, I used his denial and his own management right clause for our argument and won. Retroactive back pay at time and a half. That is an example of how our grievance procedure worked.

Sometimes it even became necessary to agree with the employer. You had to be honest; not all workers are right. A good union shows that they mean what they say about providing better health-care relationship. This was something that the National Union also never thought of.

CHAPTER 32

Union Getting Larger

As my time continued with our union, it did get larger and much more cooperation with our organizing efforts to Brockton, Barnstable County Hospital, and Falmouth, Cape Cod, and Martha's Vineyard Hospital, where I quickly found out that they had their own code. Not to be grouped in with the Cape All Nursing Homes mainly because they were nonprofessional employees, and it became much easier to negotiate contracts. I was still receiving some infighting, mainly because some of the New York organizing people said that they were supposed to be doing the negotiations. It was too long, I believe in 1974, at a National Union executive board meeting I brought up the New York organizing problems with their people, telling them if they wanted to keep my cooperation, they would have to rein in their people. Or I, as the president of the Massachusetts Union and a vice president of the National Union Executive Board member of this AFI/ CIO, I would file an official complaint, something that they thought I wasn't aware of that I could do as a member of this board. It reminds me of that later meeting to draw up our first union bylaws when the president (Leon Davis) said, "I intimidate people because I think it was that I thought things out before casting my vote and asked questions." Also we were in the middle of writing up those bylaws. We were having meetings in Pauling's New York, where I was with other district health-care union officers like Doris Turner and Henry Nicoles. I had a much different approach to things about our union than anyone else. I asked questions that they either weren't interested in or knew and didn't care. That wasn't my way. Such as when we have a delegates' meeting in New York. How were they chosen? Because if they weren't voted in by the members they represented, then I thought that this was illegal because they were given the authority by us to make decisions. It would be invalid. Everyone was alarmed by that. I was well aware of the coming bylaws that would allow the National Union to raise dues on members without their involvement. So you see, I had cause to be concerned.

I still wanted to get to the bottom of what was going on with our union dues, more so because I was in a position to ask and receive answers. The more I was being stalled from seeing the books, the more I got suspicious. I noticed Bob Mulinkamp, who was always at our national board meetings, acting like he ran errands for Leon Davis. This immediately raised my suspicions. As time went on, you will see how my suspicions were right.

Much later, things between the National Union and me came to a head. During that year, the New York team were forced to leave, and my two original coworkers who started our union stayed, although the irritation was still alive.

The very next year in 1975, we had enough dues-paying members, about 1,500, which gave us enough to become the National Union of Hospital and Health Care Employee Union District 1199 Massachusetts. That meant that we had a certain amount of independents, although we had to send our monthly union dues to New York before being sent our percentage. This was the way the National Union still controls the district. I will go into details shortly. But now we were totally in charge of our contracts

CHAPTER 33

District's First Election, 1975

Elliot Smalls said he was leaving for personal reasons. I was thinking it was because the National Union that had sent him here in the first place was finished with his environment, and it was time for our first elections for union office. I was elected president, and Judy Cooper, who worked at Jewish Memorial Hospital, was elected vice president, and Ms. Westmore was elected secretary-treasurer. Zelma Thorton didn't want any official position outside of being an executive board member

The two original organizers ran for office also but didn't do well. That didn't help our relationship.

The very first thing I did was move our office to the tenth floor, where we had more space. A private office for me, a private office for the organizers to meet with potential new members, a workroom where we printed our own monthly newsletter because we couldn't rely on the National Union to always print our good news about union victories, and grievance won. Finally, we had a large space where our secretary worked and files were kept, finally giving everyone a raise. Everyone was happy for the moment at least. We even had use of the sixth-floor meeting hall. So I could have use of a bigger place to hold monthly shop stewards meeting. This was one of the few very important times for us. It made us all feel like we had come a long way. This idea came by way of our district executive board's first. (Appendix 11 will show a picture of me as district president and a picture of me attending am AFL/CIO executive board meeting.)

As time continues with our union, it did get larger because the district executive board put them on notice: do your job, or you can go. Cooperation with the staff improved our organizing efforts to Brockton, Barnstable County, Hospital, Falmouth, nursing homes near the Cape, including Martha Vineyard. Hospital All Nursing Homes mainly because they were nonprofessional employees, and it became much easier to negotiate contracts, in part because of our reputation of being a positive influence. However, my personal time with my family started to take effect. Not so much that you could see it at first. But much later, you will see.

CHAPTER 34

University Hospital

University Hospital is now, I believe, called Boston University Hospital Center. It was organized by my two organizers and Ms. Dianne Dean, who worked there and was a member of our district executive board. This was our biggest hospital union contract in 1975, as I recall. I started my set routine of gathering all information about the hospital and getting to know and explain to the members what contract negotiation was. I had meetings with all three shifts across the street at a church, where I gained the help of union members from all departments to join in preparing their contract request and sitting in on negotiations. This was the best way to educate and allow them to be a part of the new union contract that they would have to honor. This was always very important—both parties honoring what they agreed to. This is always why I showed the importance of a great management rights clause. "Just cause" being the most important words. I had to keep telling this to all three shifts members in order to make them understand that, without it, a contract was useless. Our first contract included a serious wage increase with the various nonprofessional debt housekeepers, dietary, secretary, nursing aides, and all other employees that couldn't hire or fire; that was how all our contracts were—getting an increase in wages that were more than $9 per hour, vacation about the same as Jewish Memorial Hospital, and any other benefits that you had before the union contract, which also included the ability to retain your pension plan. This covered more than 350 workers where our only other hospital of any size covered 200 workers: Jewish Memorial Hospital, our biggest hospital at that time. So you know it had to be done right the first time.

This was a two-year contract that, in the first year, all members received a $0.50 cent an hour increase every six months and in the second $0.75 cents an hour, giving all current members a base pay of about $11.00 per hour. Plus, for the first time, the ability to apply for an upper-level job with training provided even if they would be out of the Union. In all of our union health-care facilities, we had our own bulletin board. I perhaps mentioned it earlier, but it's worth repeating. I even recall the name of the representative of the hospital: a Mr. Bluementhal. We later became good business friends even though we were on different sides. Even later I was invited to be the featured guest speaker at the Massachusetts Hospital Association meeting in 1979 (Appendix 8 will show a letter from the association).

This was something that had never happened before at any management meeting. I believe the main reason is because they didn't know much about us, and I had a reputation of being honest and fair, proving that with our being there, people were much more inclined to stay, and the working relationship was much better. This doesn't mean that their doors swung wide open, but it did make us look like the coming of the devil.

CHAPTER 35

Asking Too Many Questions

The National Union's primary interest was always about collecting dues, never about the interest of the members. It wasn't enough that all union dues was sent to New York first, then after they had taken out their portion, said to cover financial concern that I always questioned, the rest was sent to the district. I seriously disagree with this, mainly because we didn't have a monthly report on how much dues were collected and what was spent for National Union concerns from the results.

I came to that conclusion after I demanded as an executive board member to see the financial reports. All the district members were getting subsidy pay from the National Union to increase their pay. I knew this had to be true because at a delegates' meeting in New York, I had my delegates an executive board, meningeal with members of the other members and find out such things as what they pay in dues, how many members they had, how many staff members, including to level people, and any other information. At general meetings like this where everyone was really there for a great time, everyone liked to brag about their union. You would be surprised what this information can do to help you determine what that district or area is spending in the service of, something that I had learned from my grandma. Always find out as much as you can before the doors are shut. When you become a district, all experiences had to be paid by you, and that meant that these officers had to be subsidized by the National Union in order to receive the kind of paychecks they were receiving.

I hope you are getting a much better picture of what I am talking about. I knew I was alone in this effort, but it didn't mean that I had to just fall into line. Remember my grandma instilled in me the meaning of self-respect. Now it was being put to the test.

This was just one of my differences with the National Union. Bob Mulinkamp would still show up in my district unannounced—no appointment, talking to the organizers, without my say-so about things I wasn't aware of. I brought it up at one of our National Union board meetings, and the other members joined me in my complaint, something that they have been aware of but never said anything. However, in my case, they agreed to have him call for an appointment before coming and say first what he was coming to talk about. You may wonder how I was able to get anything done for our union members. With so much opposition, all I can say is that I have always believed that if you are doing the right thing, then you will be of service at least for a while. Me and my executive board members were involved in many of the National Union's meeting. It wasn't normal, but I had insisted, saying they were also elected by the dues-paying members and therefore entitled to be here. I said that district executive board members are a coequal part of the representation of their district, and they may not have the authority to vote. They do have a say in what is discussed. However, if the voting official can't be present, then an executive board member may vote in his absence. It was discussed, voted on, and I won.

On some occasions, I must admit that driving back to Boston was never relaxed. This was part of my displeasure of having Bob Mulinkamp coming to my district and eventually told him what I thought to his face. I had learned from childhood that if you have a problem, you correct it by confronting it head-on. In this way, he knew that I was watching him even if it wasn't true. You must understand in any union, there is always something going on behind your back, and it always has something to do with money.

Members must always be actively involved in the operation of their union. It's your right and your responsibility if you want to keep your union something that you will be proud to be a part of. For example, look at what has happened with the president of district 1199 Massachusetts. SEIU Teryek Lee in 2017 being charged with crime is an insult to my former union that I helped found. It hasn't been talked about too much, which is good for our union members, but was published in the Boston Globe in 2017 not too long ago. I will talk more about this later.

My point is, he was published to be the first black president of this district. The truth is, he wasn't even born when our union came into being.

I am a serious union supporter but deeply believe that all unions live or die by their leadership, and the information that the members have a right to know such things in general affect the union. Like anything to do with funding pension plan and other benefits, it isn't as hard as some want you to think to understand where your union dues are going. If you can count your paycheck expenses, then you can understand what and where your union dues are going. If, for no other reason, it keeps your union officials honest. I say this from personal experience and know that the practice of stealing never changes just the people do as they reach retirement.

CHAPTER 36

Homelife

I want to spend a moment to tell you what was going on in my family life during all of my union efforts. I traded my 1966 T-Bird for a 1969 Ford station wagon. Married with one daughter, it wasn't too long after another was on the way in 1972. She lived on Dennison Street right next door to sad Sam Jones, whom I told you played for the Boston Celtics (number 24). After we married, we lived on, I believe, Bancroft Street off Cummings Highway, Mattapan. This happened while I was at Jewish Memorial Hospital, driving, and union negotiations activities. My wife was a beautiful five-foot-seven lady that eventually captured my heart and soul. She worked at the hospital, and we dated regularly but understood that we weren't going together or at least that's how I knew it to be. In 1969, we married while on vacation in Barbados. My daughter Demeta was born on October 13, 1968. By the way, let me confess: I was the one with the stomach pains. So bad I went to the hospital. They couldn't find anything wrong, but when they took my temperature, the nurse asked me if my wife was pregnant; I said yes. She laughed and said I was suffering from morning sickness. The problem was at that time we weren't married yet; I wouldn't wish this on my worst enemy. Before we got married, I do recall I wasn't quite ready for marriage, not until Zelma Thorton convinced me that I would regret it when I see her going down the street with another man. That was enough for me. She was just like my grandma in how she expressed her thoughts. I think she was born about 1895.

My wife didn't have to work after my first daughter was just beginning school. I was getting $125 subsidy, so I was making enough money that helped to support us. I always brought home my paycheck so my wife could handle all the household bills. I would come home mostly by 5:00 p.m., unless some union matter came up. Most of the time, I would cook because my wife wasn't good at cooking at that time what I called real food. She cooked things like veggie hamburgers and the like. My daughter and even my wife enjoyed meat and potatoes along with veggies. I would prepare such things as meat loaf, made with sausage, bacon strips, and tomato sauce; chicken and beef stew. On some holidays, I would cook a turkey or honey ham. All with fresh vegetables. Sometimes we went to my aunt's house for dinner where she would always have my deep-dish lasagna. In a couple of occasions, we drove to Toledo to enjoy the holiday with my parents. I want to say in time my wife got the hang of it, thank God. It was during this time that I taught her how to drive. You see, there was no driving school yet.

This got even better when my second daughter, Leah, was born on November 8, 1972. Now with three women to cook for in the house, I had to be careful in sharing my love, three ways because my wife was as important as my two daughters. I was very good at doing things like changing diapers the old way, the three-corner type putting the soiled one in the diaper pail. This started with my firstborn. I thought that the only time my daughter's spoiled their diapers was when I was home. Plus, when it was always time for me to take a bath or use the toilet. I would ask if anyone needed to come in, and no one would say they needed to. But more often than not, one or both of my daughters would just come in; and if I was in the tub, they would just jump in. Thank God it was usually filled with bubbles, but it wouldn't have mattered. They are your heart and soul. It's funny when you get old how clear your mind is about certain things. I was very good at washing and braiding their hair. They wanted me to do it because I could dry their hair without pulling their head. My wife was a little upset about that because they would always tell her, "I'll wait for Dad to do it."

You know deep down I enjoyed all of it. On Friday night when I would be watching TV, they would always curl up on me, with my youngest always with her army blanket that she took everywhere like Peanuts in the funny papers, on my lap, waiting for me to doze off. This will always be the case. Then one of them would always turn the channel. When I would wake up and ask what happened, they would always say my program was over, so I could go back to sleep. I always knew it wasn't true, but you know what you will do to keep your daughters happy. Oh yes, my oldest girl learned how to make me my one Captain Morgan rum and Coke, and then my youngest would always bring it to me in the living room, of course spilling some on the way, but who cares? You're being taken care of right. Their play time, thank God, was generally in the backyard with our owners' two girls who were the same age as our daughters. However, when I was home, my girls would always come in often to make sure I hadn't gone out. If they caught me, they would always demand to go with me, especially on the weekend. Now looking back, I wonder why it's always so important for girls, always wanting to go everywhere I go. Was my life that interesting?

Many times, my wife would be free to go out and enjoy herself. Or we'd get a babysitter at least once a month and go to a jazz concert or to Berkeley Performance Center to see one of the R and B musicians, like Chaka Kahn, Commodores, or Teddy Pendergrass. This is how it was in my house.

During the school week, I would sit down with them while they did their homework, making sure that math was done the old-fashioned way without the new calculators. My wife was also very involved. I think we both wanted them to think for themselves. Something that they learn too well because they became just like me.

On the weekend, I would take the clothes to the laundry with my daughters to give my wife time to sleep late and let my children enjoy their weekend hot dog and french fries with a drink. That was next door to the Laundromat off Cummings Highway. I think that's the real reason why they came.

This was an ongoing event where they would put their mother's bra over their head and spell out Mickey Mouse, making me embarrassed. We. would go to the drive-in. Once, my daughters convince me to take them to see The Exorcist. I had warned them about it, but they said their friends had seen it, and it was funny, so me and my wife decided to take them. Well, in about ten minutes into the picture, my daughters screamed and got under their blankets and, for what seemed about the next two weeks, slept on my side of the bed. My wife wouldn't move over, punishing me for taking them, not taking any responsibility. Oh well, it was kinda my fault. It was always my fault when my wife thought or knew that they had done something she didn't agree with.

God had punished me by making my daughters look and act just like me. I feel that way today. They always said even to this day, "Dad, we see you do it, so we act just like you." What can you do or say? They are your daughters, and if they say they want to be just like you, aren't you going to love that? A true father will always spoil his daughters.

When they started going to school in the METCO program, this was during the time of the ongoing trouble (see Appendix 8) with bussing students to all-white schools, mostly in South Boston. I would take them to Simcoe on the Bridge, a familiar hot-dog place in Mattapan, Massachusetts. I would always give them a Twinkies or something for lunch—outside the house, of course—before going to work. Come to think about it, I can't remember her ever being the one who offered up the treats. I guess that's what women do so they can always blame the man if the children get a stomachache.

Their mother would pick them up. I recall one afternoon when I came home, and my wife got on my cast for giving them cake for school. When my daughters came into the room, they said that I made them take it. I was double-crossed. However, it didn't stop there. When we went grocery shopping, I used to put my youngest daughter on my back in a child's backpack as we would complete our shopping, and the counter employee would total up our bill. She would say, "What about that up there?" pointing to my shoulder. I would look back, and my youngest daughter would be eating and sharing what they had, so I was forced to buy it. When we got home, I would try to hide it before the boss found out, but my children would always get me in trouble. She would say more often than not, "You never discipline these girls. They're just like you." I would turn away and say softly that I was good with that.

Once, I came home, and they had washed all my pipes. At that time, I was a pipe smoker, at least up to that moment. When I got home, my wife told me what they had done, and they were in their room waiting for me to scold them. Well, you know what happened with that, right? I whispered and told them to yell while I hit the bed. When I was finished and they came out rubbing their backsides, my wife stood there and said, "You spanked them, right?" I just looked at her and said, "Isn't it time for dinner?" Now don't think I didn't discipline them. They followed my rules to the letter. I must say I can't think of one time that they disappointed me.

I remember one important time when they didn't bring home a notice from school saying that they were getting out early tomorrow. When I came to pick them up at their usual time, the lady that owned the beauty salon there next to Simcos said they had been waiting since twelve thirty, and she asked them if they wanted to wait inside. But the oldest said their dad had told them to never go with strangers. She also said, "Your daughters are well behaved." I was first in shock but greatly relieved. When I came back down to earth, I asked what happened to the notice. Of course, each one blamed the other for not mentioning it. That's life being a caring parent, something I know every parent experiences. It like a rite of way or something.

Sometimes when I picked them up after school I would take them with me to meetings. I had them sit in a corner and do their homework. I didn't believe in letting them use a calculator.

All these things occurred when they started going to school in the METCO program in Lexington, Massachusetts. This was during the time of the ongoing trouble with busing students to all-white schools mostly in South Boston. I would take them to Simcoe on the Bridge, a familiar hot-dog place in Mattapan, Massachusetts. I would always give them a Twinkies or something for lunch—outside the house, of course.

This was the same place that everyone went to in the summertime. You could buy anything from all kinds of ice cream to fried clams. When we went there, my wife never complained. I wonder why.

KENNETH WALDEN

A very strange thing occurred. I was in New York for a meeting. It was 1975. I was attending a National Union AFL/CIO executive board meeting, and all of a sudden I felt the need to call home. I asked to speak to the girls, and my wife said that Demeta was complaining about a stomachache. She thought it was because I had given them something sweet before I left. I knew that I hadn't. So when I spoke to my oldest, she said she was having stomach pains. I asked my wife to take her to the hospital, but she said it was nothing serious. I felt that I should go home, and it was the right decision. I rushed my daughter to Boston City Hospital, and they found that she was having appendicitis, and if I hadn't brought her in now, she would have died of appendicitis. I have been blessed with these feelings all throughout my life. I have always thanked God and my grandma for His blessings.

Another event that stands out is when we took a trip to Toledo to spend Thanksgiving with my family that same year. When we were ready to come home, my father drove us to Detroit where most travelers went instead of trying to fly into Toledo. The highway became icy, so I told my father to slow down because I didn't feel like taking that flight. My family knew that on occasion, I would get a bad feeling about something, so he didn't say anything. He listened, and we took another flight. The flight we were supposed to be on landed at LaGuardia Airport, and when the passengers got off and walked past the lockers that were there during this time, the lockers exploded, and many were killed. During those days, you changed planes at LaGuardia for Boston. Lockers were along the wall for the convenience of the travelers. You know I thanked God for his blessings I remembered that it was my second time. being close to death by way of an Airplane (Appendix 5 will show details). This was another one of my heart feeling that something was wrong.

144

I had some out-of-the-house responsibilities also. When my aunt retired in 1976 from contracting type 2 diabetes, it was up to me to administer her daily insulin. No one else would do it except Joan, whenever you could get ahold of her. My uncle died after sitting in the bathroom. I found him sitting on the toilet, dead. Not long after that, we—meaning my wife and I—had to go to Washington, DC, because my aunt's health was failing. She died shortly after she came home. She had left in her will money for Karen. Karen wanted to buy a car, but her mother was against it. So wouldn't you know, I was called in to decide. Of course, you know whom I sided with. I said, "Karen is old enough to handle her own money, so let her decide." You know that Karen was happy about that. However, my reason was personal. I figured it would give me the freedom I so desperately needed from always being called to pick her up at one place or another. Before I move on, it didn't fully work out that way.

One night about 2:00 a.m., I was called by my aunt. She said that Karen was stranded on the highway near Brockton, Massachusetts. So you know what happened next—of course, not without some complaining from my wife. We got there and picked her and her two girlfriends. She knew I wasn't pleased and that my wife had complained. So she did her usual smile and said, "Who do you love more than your cousin? You wouldn't want me out here defenseless." I said, "There's no one I know that would come close to you." Her girlfriends said that she said I would come, and they had heard a lot about me. You know how young teenagers talk.

On the other hand, my wife was so close to her mother that when I brought my paycheck home every week, her mother always had some form of financial problem, and she would spend all of her free time on the phone with her even when I would be trying to reach her. Back then, phones were a lot different. When you called someone and couldn't reach them because they were on the phone, all you got was a busy signal. The only other choice you had was to call the phone company to ask them to interrupt the call, something that I was forced to do on occasion. Plus, in 1976, her mother had gotten pregnant. I was asked to fill in as her husband at the hospital because her mother didn't want to be embarrassed. Needed to tell you that for later in this period.

My aunt always had some kind of summer cookout. We were always invited. But my cousin and aunt always noticed that my wife and mother-in-law were never that friendly. This became serious later on.

I do recall on one occasion my cousin who is Joan's sister came to our aunt's house for a cookout. I noticed bruises on her arms and asked her about it. She won't respond. But our aunt said something like, "I guess he still abusing you." I said, "Who are you talking about?" Our aunt said, "Her husband." Well, you know what happened next. I left the cookout with my cousin and aunt with me. I went to her house, got my tire wrench out of the trunk of my car, went in, saw him half-drunk and saying to my cousin, "Who told you that you could go out?" Hearing those words did it. I broke both of his kneecaps and told him never come back because this was nothing compared to what will happen. My aunt wasn't too surprised, but my cousin almost fainted. But her problem was solved. When we got back to the cookout, Karen had just come, and she said, "I guess I missed all the excitement." My cousin was afraid to go home, so she stayed at our aunt's house for a couple of days until she calmed down. Of course, she said, "I wish you hadn't found out because you always go too far." Our aunt said, "That's what he's here for." My wife and her mother stayed confused because we didn't tell them exactly what happened. Just one of those things, I guess. It's like if you are going to do something, just do it. No time to discuss it.

She called an ambulance, and when they came, two police officers came and asked what had happened. I told them that her husband was beating my cousin up to the point she was afraid to say anything. "Take a look at her arms." Well, don't you know they asked her if she wanted to file charges. She said no. The police officers said, "Well, you must be glad you have a relative that loves you," and left. Back then, they didn't care too much about blacks fighting blacks anyway. I can't remember what happened to him, but I do know she didn't have any more trouble. I always had this dark side; it must be the devil in me.

By the way, I want you to know about two lucky events. First when I went to the store about every other morning for something, including my wife's personal monthly supplies. You know what I mean. I think she did this on purpose. Anyway, one morning I went, of course with both of my children, who always waited at the door, wanting to go with me, only because they knew they could get me to buy something for them to take to school. Of course, I would do it even if they were caught, they would betray me.

On one occasion, I went to the corner store on Cummings Highway as I usually had to do at least three times a week to get something like milk, bread, and even my wife's first-of-the-month item, the ones with the yellow stripe. I believe she did that intentionally. My daughters always had to go with me and asked me to try one of the state's new scratch tickets. Well, I did, and wouldn't you know it, we won $500! Needless to say, I brought it right home and gave it to my wife. Another time, I went to Wonderland, a Greyhound racetrack, for the first and only time. I had just $10. I had gone with some friends that I still played a little basketball with. I would often have my family there watching me play. Well, wouldn't you know it, I played the trifecta in the second race, a race where you picked the first three dogs that came across the finish line for $3. Well, I won about $3,500. You know, I didn't play another race. I even took a check so I wouldn't spend any of it. I used the other $7 for a hot dog and a Coke. That Christmas was a great one for all of us. You know that my wife included her mother in the benefit. I wanted to put her on my income tax return. However, her mother did have one spoiled sister and three younger brothers to feed, I guess having some kind of trouble with her husband, something I stayed far away from.

I often came back home during the day to have lunch with my wife of course Desert Also. I began to wonder why we didn't have more Children, but I guess we were lucky.

When we went shopping at the Quincy bargain center, my aunt and sometimes Karen included, we always found great bargains. This was a regular event. This for me and my children was a great day. It was like a family outing. Not always the case for my wife. During this time, I must admit, things got very difficult for me to be home, at least in my wife's point of view. This was 1977, a very memorable year for me.

One very important thing that started to come to mind was the reality of the fact that my life was becoming much like my father's. I'm talking about the obligation of family and the obligation of his job. As we grew in numbers, I had to spend more time away from home. I really didn't notice it at that time. I thought I was spending enough time with my family. At least my children showed no effects. Not that I was fully in agreement. But must be honest and admit that I hadn't begun thinking about it.

It's strange how when you have gotten much older and look back, there is much you see and understand that you couldn't see before. I truly believed that my grandma was causing me to see life a lot differently. She hadn't lost her place in my heart and soul. She always said living is an education in itself.

We were living very comfortable as best you could during those days. Plus, my girls were spending time at little girls' sleepovers in Lexington, Massachusetts. Remember, this was the 1970s. The busing problem was still causing a lot of unrest. Thank God for the METCO program. By this time, my wife had been going to Simmons College. I believe she was studying journalism. So it sometimes meant, if I had to work late, that either her mother or on occasion my aunt or cousin Karen would babysit. We also were able to have a second car.

CHAPTER 37

Life Beginning to Change

N ow at this time in my life, I could begin to see much more about trying to do better for your family. I spent many years blaming my Dad for not being there for me but couldn't see the reasons. Then as I see it much clearer, it was much more difficult back then, and only if you were lucky as I was to have grandparents to assist. There is no telling where I would be. I still had much to learn, and it was painful.

My homelife, as I saw it, was quite well, considering my not being home at my usual time. My wife had her own car. I had several years earlier taught her how to drive. We were financially okay.

The job was getting a lot more difficult to handle, but we were getting bigger, especially with the nursing homes. It appeared that they were calling us to come. One big problem I had was sometimes I had to go to Martha Vineyard. Many times I couldn't get off the island until the next day. As time went by, I believe my wife was beginning to think like something else was happening, although she had known about the problems going there in the past. I should have read between the lines. Because by December, things came to a head. On

December 17, 1977, my life with my wife and family came to an abrupt halt. A day you never forget.

That night, I was home enjoying my family and the personal company of my wife. When all of a sudden, there was a serious knock at the front door. I got up to answer, and to my great surprise, there were two white police officers asking if I was Kenneth Walden. I said yes; then they told me that my wife had filed a restraining order against me, and I would have to leave immediately. During that time, there wasn't a damn thing you could do about it. I didn't get a chance to say goodbye to my daughters or ask my wife what was going on.

The only good thing was, my daughters were asleep. I had to get dressed and leave immediately, not even knowing where I was going to go. I went to my aunt's house, and I don't need to explain what took place. The next day, I called her because I at least needed my clothes. So I went to get them with Karen because I felt I needed a witness. I was still in shock and didn't trust myself or my wife. However, I had trouble keeping Karen from jumping on her. My children had already gone to school. After a few days, we met and talked. She wouldn't or couldn't say what brought us to this point. However, here we were. She said she was sorry but had met someone else and wanted a divorce. I couldn't believe what I was hearing I thought I would wake up from this nightmare at any time, but it was cold-blooded real. To spare myself some pain, I will get right to the courtroom divorce.

The judge just so happened to be the same judge that heard my first divorce case. When she proceeded, she turned to me and said, "I believe I know you, right?" I said, "I don't think so." But she responded, "Oh yes, you were that nice service member that was here several years ago. So what's wrong this time? You seemed like a decent young man." Then she turned to my wife's lawyer, who just happened to be the man that I stood in for when her mother was having his baby. My aunt was there and, of course, told the judge all about it. The judge remembered her also and responded by saying, "She is a good lawyer." When the judge questioned my wife about her reason for the request, my wife just said she wanted out of the marriage. The judge asked her how long were we married, and had I been a good husband and father? She told the judge we had been married for about seven years, and I have always been a good husband and father.

So with that answer, the judge said, "If you really want the divorce, I will grant it, but I will not put any support amount attached to this order. Because with the kind of attorney you have, I am inclined to believe you are being ill informed. But that is something that you will have to live with. Regarding the children, the father will be allowed to see them during the school week if appropriate and every weekend unless you move out of the state and he signs the papers."

Then the judge turned to me and said, "I'm sorry for your situation. You appear to make bad choices, but you have God on your side." My aunt said, "Thank you for everything," and we left. I felt like I had just lost my grandma again, but my aunt explaining to me that I had just gotten a big break from a judge that is known not to ever give a man a break, and this was my second time.

If there was ever a time I could use my grandma, it was right then. But what I had was well-intentioned relatives that wanted to take some form of revenge. Knowing that in this state and time, I would go right to jail without a trial. More important was, there were my daughters to think about. I did have a big problem controlling my temper. Looking back, I'm glad I did so.

What made it even harder: the man she met and married one week after our divorce was Kenneth Davis, a friend that I played basketball with just a couple of weeks earlier. He was in charge of the new YMCA on the corner of Millenia

Cast Boulevard and Warren Street. Now that's a real test of control, wouldn't you think? And I want to point out that it's been more than forty years, and it still feels like yesterday. Not an easy pill to swallow when you are the victim.

After the 1977 school session, I regularly saw my children and was even asked to represent the family at parent-teachers' meetings and some happy moments. I was forced to decide to let my ex-wife take my children with her to Chicago where she and her new husband planned to live. I don't need or want to explain the pain and suffering that both me and my daughter went through during that time.

I had no way of proving that I could provide a home for them when I was tied up with so many union problems. It wasn't easy for me or my daughters to say goodbye, but I knew that they wouldn't forget me, and it wasn't too long before I was proven right. My children and I had a kind of code when we talked over the phone. If everything was all right, when I asked them if they were still eating catch up on their eggs. If things were okay, they would say yes; but if not, they would say no. This is because they eat their eggs the same way I did. Margo would always call me to speak to them because they were just like me showing that they didn't like what was happening. Of course, I took this as a positive sign and did nothing to curve it.

Margo and her new husband had the nerve to stop by my family's home in Toledo, Ohio, while on their way to live in Chicago. My mother and father just took it in stride as if they didn't care. I didn't know this until my brother told me about it sometime later. When I asked why they let it happen, my parents just said they were thinking of the grandchildren. I guess that's an answer, but not one that I expected. I have never been able to forget or forgive my ex-wife because not too long afterward, when they broke up, she was calling my family, telling them that I wasn't sending any support money, and my parents apparently believe her and sent money. I found this out from my brother also.

I sent my mom check recipes that proved that she had been lying. But it had little effect on their attitude.

My daughters always called me at least two-times a month, and every time they wanted something, and their mother would say she could afford it. I would always send the money. This I did for both of them straight up through college. No matter if I was going through good times or not and there were many bad times coming. I never faulted on my financial responsibility. Let me tell you that even today, when I think about them enough, I can hear their voices and smell their hair. If you are dedicated to your children, you know this is possible, right? I want to say, if it wasn't for Zelma Thorton, I wouldn't have handled this as well. I can't pretend that I handled it with grace. But with her substituting for my grandma, I was able to hold it together.

Needless to say, this is as hard to write as it is to remember. But I can tell you that I have never used or even told my children about the details. What should always be the main concern for your children. They deserve the love, respect, and protection of both parents. Because they are God's greatest gift.

CHAPTER 38

Union Troubles Get Worse

Here I am, Christmastime, caught between a rock and a hard place, enduring the pain of the loss of my children and the pressure of staying on top of everything pertaining to our union.

The National Union quickly picked up on my home problems and didn't waste any time mentioning it, saying very cleverly, "I am sorry about your divorce. I guess it's the stress of this job." I would never respond because I knew to do so would allow them to continue to mention it.

We the National Union was working on our first union bylaws, sill having to present it to the entire delegates for a vote of approval. There was still the problem of how and who had the authority to increase union dues. I was continuing to hold on to my view that it was up to the members affected. I was trying to at least get the districts excited because we were supposed to be in charge of our own membership. Call it negotiations with ourselves who are supposed to be the best at it. This was the only weapon I had. Not even those of us who had districts spoke up on behalf of this claim.

You would think that I was negotiating with a hospital for the first time, trying to take money from them. There were things that I was pleased that the board and I came to an agreement on: educating shop stewards so they could handle an arbitration case. They saw the legal savings on this. Monthly classes for shop stewards to learn the details of their contracts.

However, getting them to understand that improving the union's image in order to encourage a potential healthcare facility to see the benefits of the union for the good of the patents was more than they could handle. This was more than they could envision. All these things came from what I had already begun to accomplish. But as you can expect, I was given no credit for those ideas.

At home, I was more stretched out than ever. Meetings from Martha Vineyard, to Salem, Massachusetts, couldn't use my two longtime organizers to service the contract in any facility because they either weren't interested or couldn't, and we couldn't afford any mistakes. We were still beginning, and of course, the National Union was just waiting for us to fail.

The National Union had a monthly magazine that advertised the ongoing growth of our union in general. In the beginning, we were always in it when they signed all contracts. But after we became a district with me in charge, we were in the magazine, much less almost to the extent of not being written about at all. I, of course, complained about this, and the answer I got was that it was up to the district to provide their own publication. Who the hell believed that? I immediately took it up with the entire AFL/ CIO executive board, and they said the National Union was wrong and must correct their error. They did, but it didn't last too long.

Among the general membership, my home situation became well known. For the most part, the biggest problem I had was to stay clear of all the action that became available, trying my best not eat of the cake that I had cooked. Thank God I was too busy to find the time to participate, for the most part.

Another huge problem was integrating our newer union members into a union that had been majority white when there was so much racial tensions still active around. I found it extremely hard to accomplish. I had to use everything I could think of, along with the vital help of my executive board, especially Zelma Thorton, whose advice I was seeking. Once again, that old grandma support came forward. Thank God I had good sense to ask for that help, remembering that a hard head makes a soft behind.

I started to have social event in their areas bring members from Greater Boston. This gave an opportunity for members to begin to see what they had in common. Gradually, it worked. Now comes the time for my next negotiations with both Jewish Memorial Hospital and University Hospital. This was an intense time not because of the negotiations but the time and mind-set that was needed to achieve both with a good result. Thanks to the fact that I had a positive reputation. Things went without much conflict.

These contracts went to early 1980. The only thing that needed improvement were wages that was negotiated: members receiving an every-six-month increase of 50 cents an hour, which brought the basic wage to around $14 an hour. This also equaled the wages that I was receiving as president. This was very important for many reasons. Mostly, no member of my district could honestly say I was using them for my own benefit

In early 1978, our union bylaws were to be presented to the delegates for discussion and vote. I didn't figure that as the main reason for our problem at that time. For the National Union to publicize our success would threaten their attempt to get the bylaws passed as they wanted.

As usual, my two organizers complained about what they were getting paid. So I put it on the agenda for our next board meeting. This is something I always did because I knew that our secretary would always share our meeting with both our organizers and the National Union. So I wanted to always have a written record.

I had always known our secretary allegiance was always with them, and I had learned it was better to keep your enemy close. That way, you have a much better view of your situation. Our board addressed the problem and called them in for a discussion. The result was that they would get a raise with the written understanding that they report to the board monthly on their progress. They agreed. I had nothing to do with their decision and made it known in our written report. All of our meetings were recorded by our secretary. So you see why we kept her for so long.

I want to emphasize that we started not understanding anything about unions or how it operated. So must say in a few short years, we had been involved. We had come a long way in keeping our union strong and clean of any wrongdoing.

CHAPTER 39

National Union Bylaws and Outside Interest

We held the National Unions bylaws convention in early 1978 to the best of my knowledge. All the membership was represented, even though I knew that in some areas, they were handpicked. For the first part, things went well. However, the raising of the dues was a problem. I was joyfully surprised. It gave me a chance to air my view about it without interruption. I informed the membership what I felt and suggested should happen with the raising of dues. After that, the Leon Davis called a recast. That was a big mistake on his part. It was meant to be a chance for the board to come up with a satisfactory response. I want to remind you that all through this period from our first meeting in 1967 until this moment, Henry Nicoles, who was Leon Davis's right-hand man, never spoke with me. But what it did was give me and my board members a chance to convince with the membership in general about my proposal. With a lot of discussion, they agreed that the districts had the right to determine their own dues and in the areas the board couldn't raise dues by no more than $1 per year and no more than $2 if or when that area had gotten more than a 5 percent raise in that year.

The national board was very upset about this surprise move it was even much more than I had asked the Board to consider.

This had to be accepted by us as the National Union executive board. I never thought that I would ever see anything that I had proposed get accepted

Driving home with my executive board, I can tell you now I was very nervous. Of course, I didn't talk about it; but when we got home and Zema Thorton and I had a private moment, she told me that she knew what I was thinking and that she was very proud of me for being a man and standing tall. It was like my grandma rubbing the top of head. Remember, there was that ongoing fear that there was a criminal element involved. It was New York, and there was a lot of illegal union activity. When we got home, I was pretty nervous about going home alone. I am just telling the truth. Remember the golden rule: nothing good goes unpunished. That has always been the case and was well aware that my time would come. I wasn't a perfect man, but I was a man that would never give up my self-respect.

It wasn't long after that the national board took up the subject of our Providence Road Island members. I had been aware that there was organizing being done there, but that was about all. Now the discussion was if we should put more organizers there. I thought that this was interesting because I recalled how our meeting started and how quickly they moved in with so many organizers. It meant that they had discussed our area long before our meeting. That meant that my coworkers had been in touch with them for a while; plus, they had even discussed me before we met.

This gave me more pause. I began to think that I was among a den of thieves, that I wasn't wrong in my suspicions.

However, knowing this and being able to do anything about it are two very different things. Now it was more important than ever to watch my back and those on my executive board. The biggest problem with all this newfound information was how long and how involved were our cofounders. Did they help from the beginning, or were they thinking their way would be the answer? At this point, it made little difference. But I never knew what the answer was.

The discussion went on for a while, and finally we decided that it wasn't worth it; plus, the organizer who was in charge was a personal friend of Bob Mulinkamp. This slipped out during the discussion. We then talked about how we were going to handle the delivery of our bylaws to our members. It was decided that each area and district would handle their own because it would give the leadership a chance to have a discussion with their members.

CHAPTER 40

My More Than Ten Years with the Union

I had come a long way, have learned much, was able to look at and consider what had been going on around me. When I was in the office, things happened; meaning, the secretary and my organizers had changed to where you would think we were strangers. I noticed that they couldn't look me in the eye, and my cofounders stayed in the office long enough to collect their checks. On one occasion, I decided I would confront our secretary with the facts. She was caught totally off guard. I calmly told her how long and what I was aware of. She started to cry. I told her that trying to develop false tears wasn't working. So I would be bringing this matter up with the board for an answer. I then told her to take the rest of the day off to settle her nerves. I know this may sound cruel, but we were at a time when I felt we had to start cleaning house. It wasn't long before my instincts were right. We had a meeting on the subject of the entire staff, and it was decided that we let our secretary go with two weeks' severance pay. She was told this directly by the board. I think our organizers got the message. Before I go any further, I want to tell you that I was in regular contact with my daughters. That had a soothing effect on my state of mind, which was very important during this period.

CHAPTER 41

Offers Made to Me

Later in the year, Henry Nicoles asked for a meeting with me. This was the first time I have ever met with him alone; plus, most of all here in Boston. I told my board members about the meeting and wanted them to attend. We first said that our termination of our secretary must have hit a nerve. We may have been right, but we all felt it was the right thing to do. About the meeting, they thought I would get more information by meeting alone.

I gave him a time when I was available, and he said he would like to meet at the airport dining room. During those days, the airport didn't have all the security like they have today. I thought this unusual, but I agreed.

It wasn't long before we met. He began to tell me that he wants to straighten out any differences I have with the National Union, and I would be treated with much more respect, and any accommodation I require could be negotiated.

Then he began with his important request. The National Union wanted me to take Providence Island Road into our district, saying it would strengthen our district, bring in more revenue; and with my experience, the area would grow much faster. Plus, there would automatically be a sizable increase in my pay. I was shocked, but not surprised. I responded by saying, "I would need time to consider this proposal. Plus, I would have to set up a meeting with my board members, and that would take a few weeks because two are on vacation." We departed shaking hands, but I felt I needed to wash them right after. The only thing I wish for is that I had one of these new cell Phones that you can record everything.

I did go right back and tell my board members what had happened. We immediately recognized that this was a serious request. This was the first time the National Union came straight out and showed their hand. Now it was very important how we handled it. We knew that our organizers were well aware of this offer. More important, if this was accepted, it would mean that our district and I would be completely under their control. Remember I said from the beginning that it's all about the money. I was put in a position where I was damned if I did and damned if I didn't. The times were changing for good, and we knew it. The most we could do was play for time as long as we could, knowing that the National Union would be monitoring our every move. There was a good side to all of this: it was that my children were not a part of this.

CHAPTER 42

Getting Married Again

B elieve it or not, here I was, smitten by this beautiful petite younger lady whom I met during my servicing the contract at University Hospital. This was in 1978. I found myself looking for her every time I went to this hospital. I had even talked to my aunt and Zelma Thorton about her. They both gave me their approval, even Karen, who by now was a Boston police officer. This lady had a son that got attached to me very quickly. Her mother and father were very nice and old-fashioned. She had two brothers also. We got married on February 17, 1979.

We went to Toledo and had a great time with my family, who all liked her, especially my brothers and sisters. We lived in Morton Village. In Mattapan, Massachusetts, her name was Betty Dabrio, from Granada. We were very happy, but there became a serious problem. When my children called and left a message, I didn't get them; plus, she even had a problem with loving her own son. Besides this, we had no problems until the following year. I once again didn't take this as a very important problem at that time because so much was going on, and I thought I had an advantage because she had become aware of the problems with the National Union and was very supportive. We went out on occasion to the Sugar Shack and concerts that were held on the Boston Commons. We even had a couple of house parties, something that was a first for me, but it did lighten my mental load at that time. She was even able to get her own car, and she was a good cook.

As I have already said, the only thing that disturbed her was my relationship with my daughters, which you know could never change. I even tried to have a discussion with her about this problem, with no results. When she would have a temper tantrum with her seven-year-old son, which you know I had plenty of experience with, I couldn't understand why; and more importantly, she couldn't control it. I guess you want to say I made another big mistake, but you never know what you get sometimes until you bring it home. So in reality, I had myself an additional problem. But I refused to just give up. She had too much good in her and my past experience with marriage. I was in no position to judge. I've been with much worse. She did have one good side that I needed very badly at this time: it was total support. Someone I could depend upon to always be there by my side. She gave me that no matter what else happened, and things were about to get much worse.

She was afraid of what may happen to us in the future. In the summer of 1979, she began having serious stomach problems. It turned out to be ulcerative colitis. When I talked to her doctor, he said that she had been suffering from it for a while and that it can make a person extremely irritated. I began to think this was the cause of her negative reactions toward my children and her own. I talked to her mother about her illness, and she said it would happen off and on. But her daughter had always said it was nothing serious, and it always went away. The strangest thing about this was, she worked in the same hospital university where I took her when she got seriously sick. All of her friends at the hospital came to see her and said they were aware that she occasionally got sick. This meant that I was the only one who didn't know. Sounds familiar.

She did get somewhat better and was at least able to come home. But she require a kind of female care, so she required went to her mother's house. Her son went also. I tried to be there as often as I could, but she preferred that I go to work. So I did.

Soon after, I was invited to speak at the Massachusetts Hospital Association meeting. This was the first, I believe, in the nation for a union official to go into the lions' den and discuss the labor union. I thought it was a great idea; plus, it wasn't any secret that we were at different ends of the ruler concerning the workers and the effect on patients' care. I talked about in general the faults that the hospitals caused in their operations, such as how disrespectful the upper staff treated the dietary workers, the housekeeping workers, the nursing aides, and other low-wage employees. They had many questions, and at the end of the meeting, they thanked me for the information, and I said with a smile that we would talk again soon. The organizer of the event congratulated me and said, "This is a first." I left feeling that our union must be making meaningful steps if the employers wanted to talk to me. As I have said earlier, there was an appreciation letter (appendix 9 will give more information). I believe this was the reason the following March 8, 1980, I was invited to speak on live TV on WBZ channel 4 news concerning our Union. All these things we felt was seen as a threat to the National Union, meaning Leon Davis and Henry Nicoles.

I went right back to her hospital to negotiate my last two year-contract along at University and Jewish Memorial Hospital. This, I believe, was a first—negotiating a contract where your wife is a patient.

KENNETH WALDEN

It wasn't easy keeping my mind on work, the National Union, and my wife all at the same time. All important to someone in our union, it was my responsibility to protect each and every one of them.

Zelma Thorton, Dainne Dean, and Judy Cooper all chipped in and did an excellent job, saying that I taught them well. Both contract provided the members with a wage increase. Before the end of this new agreement, they would earn about $15 per hour, and that was in 1979 through 1980.

CHAPTER 43

National Union Problems Continue

At our national board meeting, knowing that the subject of Providence Road Island was going to come up, I brought my district board members. I thought this would be a sign that I had a discussion it with them. They did not know if the financial part was discussed—if, for no other reason, than to stall for time. Immediately after getting the board agenda, it changed. It was a blackout on our schedule. So at least for now, we were safe. To cover the reason why they had come, I asked that the negotiation results at University and Jewish Memorial Hospital to be put on the agenda. It was first on the agenda concerning why it wasn't in the monthly magazine. They responded that it was their error and apologized and said it would be corrected in the next month's edition. I think it was placed first and treated that way to be able to continue with the meeting normally.

With that, the board withdrew and waited for me at our hotel. A hotel was always provided for us when there was a National Union board meeting—at the National Union's cost. Such things as having a National Union celebration came up. With full agreement, and for the first time, a merger with SEIU. That was a surprise to most of us, but it was discussed by no other than Henry Nicoles leading the talk. That was enough for some of us, including Doris Turner, who had a district in New York. She, for the first time, spoke up about the problems with that move and had a talk with me afterward. She said it was a trick by Henry Nicoles with the support of Leon Davis to split up the union in order that he get a larger share. I listened but didn't say much because I was ready to trust her, remembering that she hadn't supported me in the past. I did tell her that I would talk about it with her later. All during this meeting, I couldn't stop thinking about my wife. She said she was fine and for me to stop worrying and stay focused. On the way home, we discussed what Doris Turner had told me, and Zelma Thorton said that may be why Henry Nicoles wanted us to include Providence in our district. Like my grandma, she must always be taken seriously. The others agreed.

After arriving home, we decided to focus more attention on this Providence move. However, it was time for some to take their vacations, so for the moment, it was tabled.

CHAPTER 44

Going into the Last Lap

My wife had decided not to come back home for reasons I had a difficult time understanding. However, with help from some of the women I took advice from like Zelma Thorton and my aunt, they said, "It's quite possible that she felt that she couldn't be a complete wife to you and through her ongoing support, she shows you her love as she feels comfortable giving." It didn't make a lot of sense, or maybe I didn't want to accept it. But the bottom line there was nothing I could do about it. At least this time, I didn't have to face the pain of the past. I was willing to take her back at any cost. You get married to also love through sickness as well as health. That was my view, and it never changed.

It became time for us to make our decision about Providence. We decided to say no even if it meant a serious challenge to our own future assistance. We invited Henry Nicoles to Boston to meet and this time at our office. Plus, having our organizers present to see their reaction and put them on the spot because the board wasn't satisfied with the monthly results that they were receiving. Around May, we had our meeting, which was short because we said no without giving our reasoning. We didn't feel we needed to, after all; it was one of the few decisions that we could make without the National Union interfering. Our organizers were visibly upset, and it wasn't long after that they were fired.

The big mistake I was responsible for not making was, I didn't call a general shop stewards meeting to discuss it. It cost me later.

Soon after we were met by the Labor Relations board concerning the organized firing. The board produced records showing how, why, and the length of time it took before that decision was made. The Labor Board agreed, and the complaint was dropped.

Shortly after that, my wife was visiting me when I received a call from Henry Nicoles again wanting to meet with me. He said it was very important. I said, "How soon?" He said he would fly in today and asked if I could meet at the airport because he had to go to other districts. I said, "Wait a minute," and spoke to my wife, and she said yes and that she would join me. So I said I would not tell him that I was bringing anyone with me.

I first called my board members and told them about the call and my wife would be going with me. They thought this was a good move. We arrived about lunchtime. The reason it stuck in my mind is, I remember him sitting there eating fried chicken, throwing the bones on the floor.

I introduced him to my wife, and he said, "Hello, I am glad to see you feeling better." This I found strange because I have never spoken about her condition with anyone in New York. I said, "What can I do for you because I have another meeting." He said that the National Union needed me to get the union's benefit plans into the hospital's contract. I said, "You must be kidding me. First, when was this decided? I was never at any such meeting, and you're too late because these contacts, as you know, have already been signed, sealed, and delivered. He then leaned over and said, "Don't worry, it can be done with your support, and there's a healthy financial benefit for your efforts, in the amount of $54,000 dollars." I took a second to digest what I was hearing, and my wife just smiled and said, "What do you have in your food?" I wondered where he got that figure from but never asked because once again I said, "You know that no decisions like this are made without their approval." Then we left and told him we would get back to him in a couple of weeks

I immediately got back in touch with the board. My wife told them exactly what was said. Then she had to leave, forget treatment. The board thanked her and told me that I have the best of her. I responded that I had finally come to appreciate it.

We all decided to get prepared for the National Union's big push. One thing I noticed was, they never hired our former cofounders, which told me that they were only using them; and now, if and when they came here, they would have their own staff. A case of a clean slate. I never knew what happened to them, maybe because I wasn't that interested because I have always blamed them for their greed and jealousy. If it wasn't for that, we would never have been in this situation.

It wasn't long when our recently hired secretary called me in Springfield, Massachusetts, where I was with Ms. Witmore, negotiating their fifth contract. I was told that Bob Mulinkamp was in the office, wanting to speak with me. I was first upset but did talk to him, asking what he wanted and why he didn't call first and make an appointment. He said he thought we didn't have a secretary. I said, "How do you know? And I am in Springfield. So unless you want to come here, it may be a couple of days before I get there." He said no, and he would be in touch with me. I believe he came just to see how we were servicing our membership.

I know I haven't referred to Ms. Witmore much, but I must say, she attended every meeting in Boston and New

York, always offering her short but direct opinion on all matters discussed and was a strong advocate for our district. She would either drive or come by bus, depending on how long and whether we were meeting In Boston or traveling to New York, she was always paid from our travel fund; and remember, she was our secretary-treasurer.

When I got back to Boston, I had a message from Doris Turner, who wanted me to call her as soon as possible. I was very surprised at first and knew it had to be very important if she called me. So I returned the call. She told me that Henry Nicoles, Leon Davis, and Bob Mulinkamp had met and decided to take over my district and merge it with Providence soon. Now this was no surprise, but what was, was why would she be warning me? Well, it didn't take long for me to find out. The real issue was the entire National Union was restructuring to get ready for Leon Davis to retire and Henry Nicoles to be voted the new National Union president. This would hurt her district even worse than mine. I told her thanks, and that was all because I felt that she wanted to use just to help herself in whatever she had planned because she had been part of their group for years, and she never tried to support any of our efforts.

We called a board meeting, which we all felt may be close to our last together. There was always one important item that we never had any control over was: our union dues. Without the gas to run the engine, the car can't move. We all knew that one day they would get fed up enough to try it. The issue was when the only one really affected by this action would be me. As you may recall, none of the board members were paid out of union funds, mostly because they were making as much and Zelma Thorton, Judy Cooper, and Diane Dean were make a little more than myself from the contracts negotiated at their hospital. Ms. Witmore was making a little less only because it was a nursing home. Our secretary was hired on a temporary basis, so I was the only one who would eventually be without an income. The board saw to that right away, raising my pay to a level that when I would be forced to collect unemployment, I would get a decent amount, figuring we had a few months left.

There was also the vague possibly that some form of criminal element would come forward if I caused too much trouble because I was getting hung-up calls at home; and on a couple of occasions, my tires were cut. I knew that there was no sense reporting it because this was Boston, and the bottom line was still that black people need to stay in their place. Plus, there was my wife to think about. I know you may think this is an exaggeration of the issue. Believe me, this is all the truth. You who are in this union today and enjoy the benefits have no idea the sacrifices that people made in order for you to work in a much healthier environment. As I have said from the start, union dues can be the fruit of all evil, especially when you have a pension fund involved. I was a dues-paying member of that pension fund, and they couldn't seem to find any records about my having ever been employed by the National Union of Hospital and Health Care Employee District 1199. Thank God I still retained records, and I am currently fighting for my long-overdue union pension. This will be expressed in more details later.

Just as we had thought, the National Union did hold up my dues for a month, saying there was an accounting problem. We knew better; it was a test. But we knew it wouldn't be their last. We decided filing a complaint with the National Labor Relations Board, but I said the publication of this would destroy all we had built, and the members would be the losers because the issue at this point was me, not the destruction of the union membership. I wasn't being brave, just practical. My grandma was clearly talking to me, being echoed by Zelma Thorton. Who said I was making a good point? Now

Ms. Witmore had her say, and we all listened. She said when the time came, take everything like records with you. Leave not even a pen for them because it would take them months to figure out how this district operated. This was the wisest idea that we had heard. Everyone quickly agreed. We anticipated that we had about a month, maybe two. Because we knew they would be relocating the Providence staff, and that would take about a month. Remember I had been told no good deed goes unpunished. After this, we all went out for a great dinner, like we were celebrating.

Zelma Thorton and I went to visit my wife, who was back in the hospital. Ms. Thorton comforted my wife and told her our plans and not to worry about me and my income. She then left to give us some time to ourselves. This part of my life was as painful as it was when my daughters were taken away. Going home to a lonely home wasn't anything I had ever expected. When I talked to my daughters, I never told them or their mother anything about what was going on. I always pretended that everything was fine. Even my folks didn't know. It was my rock to carry.

It was about two months later I got the message that they were coming. My monthly dues had been held up again, so I called the board to let them know. I paid out my secretary her two weeks' notice and wished her well. Then I packed up everything in the office. Thank God I had a 1969 Ford station wagon. I put everything in it and left a note saying welcome and left. The very next day, I filed for unemployment, saying that my union had been merged with another. Seeing that my name had always appeared on the monthly tax deduction, I had no trouble getting it—all before they even came and never knew. I have no idea what happened next, but much later, I was told that the union split up into several parts.

I guess you are wondering why I didn't fight to keep my place in the union. Well, all I can tell you is it was the biggest decision I had to make at that time. The reason was, we were a new growing union, and I was aware that any public unrest among us would cause great harm to our membership and undo all we had accomplished. So it was a case of not wanting to cut my nose off in spite of my face. I had talked it over at length with my executive board and, after considering all options, felt that this was a much better way to go.

Now home alone but still living comfortably, I was able to spend more time with my wife, both visiting her at the hospital or at home. I was not getting any closer to getting her to return home. I talked to my daughters more, spent more time with my aunt whose health was weakening.

My wife suggested that I take a trip home. So I did. I can't recall if I told them about what had happened at that time. I don't think I did because it really hadn't hit me yet. I do recall everyone asking about my wife. I told them that she couldn't come. Shortly after being there, we were informed that my aunt had passed away.

So I had to fly right back home and represent the family because my dad wasn't well. I returned and, with luggage in hand, went right to her wake, where my cousin held on to me like glue. For years, we spent time whenever we talked about her mother and all the joy I brought her, plus all the great times she and I had. This went on until she passed away a couple of years ago.

While living at Morton Village, it was about to become a co-op. So I got the job of becoming the building superintendent. I had to get an income; my unemployment benefits weren't going to last forever. I even had my daughters visit for two weeks, being able to swim in the pool go to the movies. We went to Simcos; now that was real fun. For those two weeks, I felt like all my troubles had gone forever.

This, of course, wasn't true; but at that moment, who cared? I even got a chance to wash their hair. I think that was a high point of my time with them. When they left, my oldest girl Demeta took my portable eight-track recorder and all my R and B recording. Now I wondered, Has she grown up that much? Of course, I didn't mind.

Time went by too fast, but we had a great time. We never talked about anything that may be happening at home. If there was, I knew that my baby girl—that's what I have always to this day call her, she would have told me in some way. They said that they drive their Mom crazy with their always wanting Catch Up on their eggs. We laughed about that. When they left the flow with children aid watching after them. They called when they got home. Their mother asked me where did Demeta get all those 8 tracks. I said from the good fairy and said goodbye.

I continued to see my wife as often as she would let me. However, I noticed that as time went on, she didn't want me to come around too much. I spoke to Zelma about this person who had just retired. She said, "It is most likely because she sees you. She is reminded about the fact that she's not able to be the kind of wife she would like to be." I asked her, "How do you know that?" She said bluntly, "I haven't asked for a divorce, has she?" I said, "I would have never thought of that." I stayed and worked at Morton Village until 1983. After, I got an apartment on the corner of Walnut Avenue and Millenia Cast Boulevard. It was right across the street from the basketball court where I had a chance to play a lot of relaxing basketball.

CHAPTER 45

Starting New Life

About this same time, I was in Grove Hall, Dorchester, Massachusetts. For reasons I can't remember, I bumped into Judge Harry Elam. He went to school with my father. He was now the chief justice of the Boston Municipal Court, the only black judge there. He happened to ask me if I was working. I replied, "Not particularly." He immediately told me to come to the courthouse tomorrow morning around nine o'clock to see him. I said I would be there, and he shook my hand then left. I went home and couldn't believe what had just happened. I couldn't sleep all night. The next morning, I think I got up at 5:00 a.m. and took my wife's son to school a bit early and was waiting for the judge at his office by 8:30 a.m., remembering never to be late for an appointment. When he arrived, he first told me that while he was chief judge of the municipal court, he want to level the playing field regarding court personnel—by bringing in more minorities

He took me down the hall to the clerk's office and told the person in charge that here was a new hire. The white manager looked at me with unfriendly eyes and said, "Welcome." My first day as a court officer/file clerk was a terrific beginning to a whole new world. I thanked God and said thanks to my grandma and began a new career. For the remainder of that year, I worked my ass off learning everything I could in order to do my job well, coming in every day at 7:45 a.m. Looking right shoes shine, shirt, and tie—all clothes that I had from my days working as the union's representative. I became friends with one of my fellow employees whose last name was Bond. I can't recall his first name and began working out with him at his home in his family's basement about three times a week. It was taking away from my life at Morton Village so much that I soon moved. My wife's son moved to his grandma's house because he was changing schools anyway, and her health problem kept her going back and forth to the hospital.

CHAPTER 46

Going to Get Along

Not long after my first six months on my new position, I began being asked to go over to the state's office building next to our courthouse at the end of the month to cash more than ten fellow officers' paychecks. This usually totaled more than $10,000. This was the end-of-the-month final payment as a result of our weekly withdrawal that we were allowed to take out of our salary. They would always call me the name Bubba as they tried to get me to believe was a sign of friendship. But I always knew it was really a sign of the way they truly felt about my being there and my being able to win the favor of the superior court judges. I never let on my feelings about this, which worked in my favor later on. Once, I was asked by my friend's last name, Bond. For some reason, I can't remember his first name. I loaned him my entire end-of-the month paycheck so he could pay his truck payment. I did it, and he said he would repay me the following month. A matter that worked in my favor later as time went on. At one of our union meetings, a serious problem developed. A longtime white member of our union who was our treasurer confessed that he had "borrowed" more than $10,000 from our union dues. As a result, the membership accepted his answer and said he could no longer be the treasurer, but he could make a monthly repayment of $300 to the union. Now came my biggest wrong decision of my entire adult life.

I was asked to be the treasurer because I worked alongside the union president, and what they didn't say opening was being the only black member, I didn't have any time to consider it, or at least that was how I remember. So I accepted the job. I started writing monthly checks for our usual bill such as our part of the health plan and what became very important monthly $1,000 checks to Ray Bolger, state senior, the brother of Whitey Bolger, the famous gangster. I never asked why, but at some point, I was told it was important because he was a team player when it came to our contract and grievance matters. I also wrote checks for meetings in South Boston, where I always declined to attend because of safety. Need I say more? I wasn't that big of a fool. About the repayment of the $300 a month, it happened for about five months. When I asked at a union meeting, I was told it had been taken care of. I didn't say anything more. I had quickly learned that using the old tried and tested rule to see nothing, say nothing, know nothing, was the best method of keeping yourself safe. Strangely enough, it was during this time my former Union got in touch with me and asked me if I still had the union files. I said yes. They said what would it cost to get them back. I laughed to myself, thinking about the two members that had been convicted of stealing union dues must play a part of the reason they needed the records. So I answered, "Two months of my overdue pay still owed me." After some negotiations, witch as they forgot I had cut my teeth on, we came to a settlement. So I hired a lawyer that I had come to know to handle the transfer. Also, I demanded that they pay for the lawyer's fee for the transfer. They had shown their weakness by agreeing so quickly to my demands.

This is something else that proved important. That would all become lifesaving things.

Right about this time, my wife was killed in a car crash. Believe me or not, but the night before it happened, I saw it and called her mother, telling her if Betty was going to New York tomorrow and to tell her to stay home; if she doesn't, please don't let her take her son. Everything I predicted came true. Her mother couldn't believe what had happened. She said that in the morning, Betty had said she was going to New York with a friend, and she was taking her son. She told Betty what I had called about, but Betty refused to listen. However, her mother was able to stop her from taking her son, who cried because he wasn't going. On the road, a truck lost a tire, and it came across the highway and directly hit the car that Betty was riding. It happened in June, Worcester, Massachusetts, 1984 of some personal business. Without telling him my reasons, I can't explain why.

The funeral was even more painful for me and her family, knowing that I had predicted what was going to happen. So afterward, I made my peace with her family. They said that they found no fault in me. I was and will always feel sad, thankful for that.

Once again I had that feeling that comes over me, and it came true. Only this time, it was personal.

Now let me return to my working relationship with my fellow Suffolk superior court officers. More often than not, I found myself writing checks or cash for things I wasn't told anything about but working alongside the president of our union, who had to cosign the checks. I wasn't concerned. I never went to any of their events, always held in South Boston. Not that I wasn't invited, but felt it was just a false invitation. My job went well during my first year for the most part. Plus, playing basketball again helped me relax.

During my second year, I was able to get Judge Robert Mulligan to allow my baby sister Mary to come to Boston and be the clerk for him. This was important for her because she was attending Indiana University studying for her law degree. That she thanked me for my help. The fun part was, she spent the summer staying with me and me doing most of the cooking. I once made what she called the drunken chicken because I cooked it with too much wine. After eating it, she had to lie down and sleep. Once she got settled, I had to cut her free enough to go out with some new friends she got to know. As a protective big brother, I had to meet them and give my approval, something she found very funny; but when I later explained the problems of Roxbury and Boston, she understood. She was used to places like Toledo, Ohio, where she grew up. Safe and peaceful. Although I was a bit too protective, I didn't admit it then. At the end of the summer, she went back to college. I soon found out that I was looked upon by my fellow officers with jealousy because of my sister clerking for the toughest judge in the court, but I got along with him. I was used to a no-nonsense person because in my own way, I was the same and understood what it meant to have order where you work, especially when you are representing the commonwealth.

During my second meeting yesterday, the president that I worked with gave his position up without telling me, and a new one was elected, someone that I didn't get along too well. His last name was Donovan. His election was partly because he had been there for over fifteen years, and I had to lead his jury when they were requested. He didn't know how, and he was embarrassed when the office signed me to head it. It wasn't my choice, but it serves as an example of how serious these longtime employees took their job because you couldn't get fired without a serious reason. Nice job if you could get it. That's one of the reasons I was disliked because

I AM WHO I AM

I had invaded the good old boys' club. Still Boston racism was alive and doing well. But as usual, I said nothing, which you know was very hard for me; but when you have no cards to play, you said nothing. I had many court views that my juries had to go to because my court mostly handled murder trials, which I can tell you is a lot different than other trials. Caution was the key thing.

Back then, there were no hand radios or a lot of help transferring prisoners from the ninth-floor dock to the tenth floor courtroom. A lot of the time, you had to use your head to convince the prisoners to cooperate. This happened to be something that I was very successful at; I can't really explain why. I can just say I never had a problem. Once, I even had Mayor White as a potential juror. But he was never chosen. I did have a chance to talk to him about senior housing that he was putting up at Egleston and Washington Station on Sever Street. I asked him what he thought about the eminent domain that happened to my grandma and was still going on. He could see that I was still very upset. He only responded by giving me the political response of, "That's too bad," so I left it at that.

I even went to schools where Judge Elam was trying to help young black people get an understanding of what they would face if they were ever forced into a criminal court. It was like me playing a defendant. I would act like a young punk on trial talking in the normal street slang, leaning back in my chair, answering by saying, "Yo, man, he tried to dishes me," and the like. He felt that if they saw themselves, they would be embarrassed. I never found out if it made any difference.

By 1985, I was pretty comfortable with my job, even taking a vacation and flying to Bermuda and staying a whole week enjoying myself. No one was with me, in case you were curious. I even took a trip to Toledo to see my family. Also

185

I had gone to Aurora, Illinois, outside of Chicago to visit my daughters because my oldest was running track, and my youngest was in a school swimming contest. That should tell you how far my relationship with my family had come.

When I returned, I was told that the president and some of the men had negotiated a contract with a dental office without knowing or even asking me if we could afford it, but this was another thing that happened to make it clear to me that I really wasn't one of them. The truth is, we couldn't, and I told the president so because we hadn't recovered almost nothing from the taken union money; and if they couldn't stop using union funds for their social events, we will never have enough. But it was like talking to a stone wall. So I did the best I could to pay for all they wanted. This included the continued monthly payment to Ray Bolger and a couple of other state house senators. This became very important later. It was during this year that the conversation around the next assistant chief was beginning to become a lot more noticeable. Even my judge Robert Mulligan said to me in private that he was submitting my name for consideration. I, of course, was overwhelmed but did not say anything. But I could feel the increasing pressure. However, I was still cashing all the end-of-the-month paychecks with no problems. I was still handling murder trials, sequestered juries, and court-ordered views, mostly having to do with men who had ties to South Boston Gangs.

To tell you the truth, I thought every now and then that the reason I was always the one that would get these assignments was that I was black, and all the other officers lived or socialized there. But it did make me nervous because you never knew what could be happening. I couldn't trust anyone. But your job is your job, and I figured, "That's what I'm getting paid for," so I always said a quick silent prayer and went on. However, I did talk about my relationship with all the other court officers, with Judge Mulligan, especially when the subject of my possible new appointment. He told me not to worry about it because he had seen me handle this situation for a while. I didn't know he was aware. He said that was one of the things that he felt I would be good for the job and the court. I did feel very good and proud of this statement because he wasn't someone that talked about things like that. It now became difficult to work with cooperation with the other court officers. However, at the end of the month, they still wanted me to go cash their checks. I guess nothing interfered with that need.

CHAPTER 47

First Municipal Court Trial:
Seventeen Counts

Now comes trouble. All things happening at once. This time, I had to gather all my strength, using everything I learned in the past.

I was charged with stealing more than $1,200 from our union fund. To tell the truth, I shouldn't have been surprised because it was about to be announced that I was going to be appointed the new assistant chief court officer. You know they weren't going to have any of that.

The complaint was filed in the Boston Municipal Court on the fourth floor of the same building where I had worked with the very judges that would hear my case. No surprise there. I couldn't think of a better place to hang a black person. Judge Elam was no longer the chief judge. His term had run out, so you know what that meant. I knew my case shouldn't be held in any court that I had any contact with the judge, but I also knew that if I was found guilty, I had a good claim for another trial. They brought forward seventeen complaints of larceny of over $100. In this way, they and I knew that If I was found guilty of just one, my Ass was grass—as if they were the lawn mower.

So the first thing I did was go to see Judge Elam, telling him just what happened. Well, wouldn't you know—he already knew and had set up a ply deal. To say I was shocked would be putting it mildly. I said, "You must be crazy! Why are you doing this without telling me first?" His response was that he was thinking of the embarrassment that it would cause my family. I said, "Family! Hell, I'm not even guilty." So I left with a bad feeling in my gut.

Now I had little time to think about how I would handle things, reviewing what I went through while at the health-care union. I figured that I had to use my past to protect my future. Before I tuned over the checkbook, I made notations of what though would be important material such as the dates and amounts of checks written to Ray Bolger and many other evidence that I could use later.

I then said goodbye to my longtime Judge Robert Mulligan, who by the way was outraged over this and wanted to get involved. But I said, "Thank you for all you have done for me and my sister. I can handle this. However, if I ever needed a recommendation, would you give me one?" He said yes and told me that if I needed his help, don't hesitate to contact him. Later, I heard he wouldn't allow any of these court officers to handle his courtroom. The very next day after I was charged, the Boston Herald and Globe had my picture and a statement that said what I was being charged with. Of course, you know it was all an attempt to get me to plead for mercy. They didn't know me well. I wasn't raised that way, and this wasn't my first dance.

Now the worst thing that happened was when Judge Elam called my family and told them what he wanted them to hear about what happened. Plus, he said he had set up a plea deal that I should take. Well, you know that's a clear violation of the court rules. So I was forced to tell them what was happening over the phone but made it clear that I wasn't giving in.

While trying to concentrate on my court battle, I also had to find a way to make some money. I have always been able to find a job because even though I was still collecting my courthouse unemployment, which they tried to prevent me from doing, I was smart enough to know you may be charged with a crime but until you are found guilty, you are as entitled as anyone else to collect. So for twenty-six weeks, I got a pretty good unemployment check of about $300 every two weeks. Not bad, but I knew it wasn't going to last forever.

My cousin Karen now had become a Boston police officer then working out of the D Street Station in South Boston. She had asked me to stay at her house, but I didn't want her involved. Plus, I was dealing with a lot of on-the-job racism, not just because she was a black person from Roxbury but more important, she was black woman who viewed her as rising above her place. Also it was Ray Bolger's town. Even by this time, we faced a lot of this kind of treatment. I wasn't a fool. I have always planned my moves ahead as best I could. That was one of the reasons I never decided to go to Toledo, Ohio. I thought I could continue to keep my apartment for a while longer because I was taught by my grandma that if for any reason you have made your bed hard, it's up to you to straighten it out. I didn't want to keep any money in the bank for obvious reasons, so any extra money I received, I let my cousin hold on to it, she being the only one I trusted. So you see, I had already made a great move even when the odds were still heavily against me.

Finally, my trial date was set, and I hired a lawyer that I knew from his coming in my courtroom for trials. I gave him my story to a point because in my situation, you can't trust anyone until you know for sure. In most situations, you would put you fate in your lawyer's hand; but in my case, I was aware that there were too many court-related people working against me.

Fortunately, my father was coming to Boston for a meeting soon; and when he arrived, we met, and I told him the whole true story and that Judge Elam had no right interfering in my case. He agreed, and surprisingly, he said, "You just be chinkie and follow your thoughts." That was the first time, I think, that I really felt I had his full support. I left feeling more confident of my decision to fight. When my trial date came and I came into court, the first thing I noticed was a black former basketball-player friend that was now a Boston police officer. He told me that he was sent here because they were expecting trouble from a defendant. I took that to be me, even though he didn't say as much. Also, I saw a black camera man that I knew from when I had been on Channel 4 WBZ TV back in March 1980. He came over to me and whispered that they were here because the court had called them because I was going to plea. I said, "No, I'm not," and he said he thought it was a setup. I went into the courtroom with all my former court officers present.

My attorney, who was paid by my parents $3,000—I didn't have that kind of money at that time—never interviewed any of the judges that I had worked for and now had a private meeting with this biased Judge Pino, who asked me how I was going to plead. I took a moment just to get everyone waiting for me to say guilty, but I said, "Not guilty, Your Honor." He almost hit the ceiling and asked me again, and I said, "Not guilty, Your Honor. I am here, ready for trial." He said, "Doesn't this man know that we have arranged for him to take a plea?" I spoke up and said, "No one asked me." The whole courtroom got upset.

The bottom line: they had to schedule a jury trial. I looked at my lawyer and gave him the finger when no one was looking. Then I told him he was fired. This would allow me the time to get a court-appointed lawyer and time for me to think about my next move because this event told me that I had been right about not trusting anyone.

But they didn't count on me knowing enough about what was happening to see their game. It was like I was the stupid nigga who didn't know anything. Well, that was just the beginning.

I want to point out what I believe is a problem with a jury trial. I remember when the time came for the court to render a decision: you were either guilty or innocent. Meaning, the court had enough evidence to prove their case, or innocent because the court didn't. Today, when you are found not guilty, it generally means that the court didn't have enough evidence to convict you. This leaves a stigma on the person being charged. Just something to think about.

CHAPTER 48

Second Trial

N ow back to my case, between the time that my case was transferred to a jury trial that I requested, where you have the option of a judge hearing your case first. Then if you don't agree with the outcome, you can request a jury trial. The problem with doing this is, although you get a chance to hear the evidence that the court has, the court has an opportunity to hear your defense; and if the judge is biased, the court can deliberately find you guilty to allow them to retry you. Just an important point. I asked for a jury trial because I put more faith in them rather than the judge (Turner) because I had seen him in action, and by law, the whole justice system shouldn't have had an opportunity to hear my case because I had a previous working exchange with him; plus, he was one of the judges that was against me being appointed in the first place. But I knew that my only remedy was to appeal later, and it may be an ace in the whole.

All these things I learned by working there and being aware learning getting educated instead of just being happy to be there. It is wise to learn more about where and what you are providing to your job than just being employed. These things a client must think about before and during your court case is being heard. A lawyer does his best work a public defender or a paid lawyer when they realize that you aren't stupid; playing stupid only works if you have no choices, and you are prepared to risk your results by some form of a surprise testimony, which in my case was the plan—mostly because the assistant district attorney had me come in and review the checkbook, which was my right with my lawyer, who never came or notified me five times over two years, and each appointment had been cancelled without him telling me, offering some form of excuse. But remember, immediately after I was discharged, I made notes of everything I could remember about the payments made. Also, I knew that Judge Robert Mulligan and a couple of other superior court judges had submitted support letters to the court on my behalf. This I was told by Scotty Rose, the past president that I worked with in the beginning. This happened the morning I came to the courthouse to get my pension funds. This was another lifesaver. He just so happened to be on that floor, although he said he would never testify in my case; he knew that I was set up. I thanked him for the information but realized that he still had to work there with his fellow white officers.

My trial started two years later in the summer of 1988. Remember. I was charged in 1986, waiting two years for a larceny complaint, not a murder trial. Remember I had seventeen charges against me. Plus, I was finding ways to support myself and send money to help my daughters, not an easy task, especially when no one would hire me with this case being held over my head. Remember, my daughters didn't know anything about this.

During all this period, I began receiving calls late at night, sometimes saying, "Watch your back," and "Bang." I knew it had something to do with my court case. One, because it didn't start happening until after my first court appearances. Plus, the only one I figured could be harmed was Ray Bolger, which meant that it had to be someone wanting to protect his interest. It reminded me of my final days with the health-care union, so you know it wasn't going to stop me. Plus, it also meant that the state didn't have a strong case, and it was always their intent to just embarrass me enough so I wouldn't get the new position. That was partially my view since they waited almost two years to bring my case forward. You would think I was being tried for killing the governor.

The day of my trial, I bumped into my former friend Bond, waiting with some of my former court officers outside the courtroom. As I passed him, I said, "Good morning," and then stated, "A real man has to do what a real man has to do." Meaning, a real man must stand up for what he believes in. I had a good jury mixed. Before the judge came, in I took this time to tell my court-appointed lawyer what he had already done wrong like asking the judges in this court to recuse himself because I have all dealt with him in the past. I was never given an opportunity to review the evidence, and we were never given a list of the witnesses against me. Also, my witnesses were never interviewed.

When I had finished, he looked at me and had nothing to say. So I finished by saying, "I will advise you on the questions you should ask." Then the judge came in, and the assistant district attorney began his talk about my wrongdoing and that he would provide proof. The clerk read the seventeen charges against me. Then my lawyer had an opportunity to give a statement, but I stopped him because I took the chance that it was up to them to prove their case, not for me to defend my innocence. The trial began by the assistant district attorney calling my former friend to tell the court what I had said just before entering the courtroom. He did, and then the attorney had no more questions. I could never understand why he was called to make that statement because all it did was to open him up to my questions. So my attorney asked him first, did he know what I meant by those remarks? He said he didn't. The jury made a snickering sound. Then I had my attorney ask how long? And, "Hadn't we been close friends? Plus, didn't I let him borrow my entire paycheck to pay his truck payment even though he still lived at home?" He hesitated and said that we both became Suffolk superior court officers at the same time, and yes, he asked me if he could borrow the money.

Finally, we asked, didn't he see my wife's son on the street and tell him that I had stolen union funds and was going to jail? He paused and then said very softly yes and stepped down.

Next the current president got up and testified that he discovered the missing money while going over the books, and the attorney said, "No more questions." I gave my attorney a list of questions first when he became president, and wasn't I kept on as the treasurer when he became president? He responded that I was appointed treasurer before he was elected, and then he said that I always keep the books, and he had to admit that it was his signature on the front of each check required before cashing. We then asked, didn't you tell him that he was being given a $50 bonus for keeping the books? He answered yes. "Then how did you have a chance to go over the books when the defendant had custody of them?" He didn't have an answer. The final question I saved for last was, I was given the job because the former court officer that had been a member for more than seven years gave up the job because he admitted to stealing more than $10,000 from the union and was told that he could pay it back at $300 a mouth.

Before he answered, the assistant district attorney objected, but the judge was forced to tell him to respond.

He said yes. Then he was asked if the officer ever repaid. He said no, and didn't the defendant talk to you about the lack of repayment? He said yes. Finally, didn't you tell him to forget it? He looked like a defeated boxer and said yes. You would think he was on trial. Finally, one of the officers that didn't have anything important to say was asked a couple of stupid questions, and then after, my attorney asked him, didn't he and all the other court officers ask me to go over to the state's office building on the last of the month for the last few years to cash their checks? And he said yes. Then my attorney asked, didn't all of you give him the name 'Bubba'?" He paused and then tried to say it was meant to be a friendly name, but the jury made a sound of disgust. That alone told me that I had begun to have the jury on my side. After that, the judge called a recess for one hour.

During that time, both lawyers met with the judge, and my lawyer said that there was an agreement that when I testified, I couldn't bring up anything pertaining to checks given to Senator Bolger. Right away, he told me how I was going to win my case. We adjourned for the day.

That evening, I went and had dinner with my cousin Karen. Then we went to Prince Hall Lodge, a Masons' after-hour dance club located on Washington Street near Grove Hall. You could only get in if you were a Mason or a direct relative of one. I was because my father and our two uncles were Masons.

The next day, just before I took the stand, I had given my attorney a list of questions I wanted him to ask. He looked surprised; however, I told him that I was aware of the judge's letter of support and that there was no one here that I had requested for my defense. He had no response.

Taking a deep breath and saying to myself, Grandma, I hope you're with me, I took the stand. My attorney started by asking me all the general questions, like, How long have

I been a Suffolk superior court officer? Where did I work, and what were my responsibilities? Was I being considered for a promotion to be an assistant chief. Did I ever join the others in any social events? I responded by saying to one of the questions. About my not socializing, it was because they were always held in South Boston, where blacks found it hard to be accepted. The jury made a sound at that point.

Now comes the good part—I had him turn to my work history. He then just asked me to explain. I said that before I was hired here, I was one of the Founders 1966 at Jewish Memorial Hospital, Roxbury, Massachusetts, of the National Union of Hospital and Health Care Employee District 1199 Massachusetts. I was the chief contract negotiator, the secretary-treasurer in 1973, a National Union executive board member a vice president of the AFL/CIO from 1973 to 1980. The first health-care member to be appointed on the American Arbitration Advisory Board from 1973 to 1980. The first president of this district in 1975. I then had him ask me about my wife passing away. I told them that my wife was killed in a Massachusetts Turnpike truck accident in 1984, and her mother had a civil suit that I took no part in.

He then asked, "When your former friend Mr. Bond spoke to your wife's son, was this after your wife had passed?" I took a moment and then said yes. He then asked, was there a union-negotiated financial settlement not too long ago? I said yes. Then he asked, was I once a close friend of Bond before all this? I said yes; that's why I was so upset that he would tell a young boy that, especially when I had given him my entire paycheck when he had no one else to turn to, not even his parents. Then he said, "That will be all."

The assistant district attorney then stepped up and first told the judge that none of the information was given before the trial. Before anyone could say anything, I spoke up and said, "You had me coming into your office more than five occasions without giving me a chance to see the material that I had turned over to you more than two years ago." The judge kept trying to tell me I was out of order, but I knew if I didn't say this so the jury could hear it, the judge would be able to rule against my testimony. So he had to back off.

The assistant district attorney got right to the point and asked about the petty cash checks that were averaging about $300 dollars. I responded and said that he should look on both sides of the checks, and he would see both signatures. Plus, in the checkbook, you can find my notation of when, why, and what the money was for—all these things that he thought I had forgotten, which brings to mind why it was important to have taken my notes early. He began to look defeated. Finally, the big question came. He asked about a particular petty cash check that my notes wrote was just after a $1,000 check made out to Ray Bolger. I said I would have to look at the checkbook in order to remember. So he handed me the book. I took my time to respond and said in a way that the jury could hear every word clearly. I remember it was additional funds to be used for a social benefit after the $1,000 paid to Ray Bolger for a new contract. That was the ongoing practice. The judge banged his gavel so hard it broke, and he called a recess. He demanded to know if I was told not to mention Mr. Bolger. I said, "I can't recall, Your Honor. I apologize if I made a mistake. I spent the night worrying about today, Your Honor." He asked my lawyer, who said he had, but perhaps he couldn't recall because he was the only one testifying. The judge called the jury in and told them to disregard my last statement. I knew they wouldn't because he had made it noticeable by his heated actions. The assistant district attorney said that was all, so the judge gave them his instructions after the assistant district attorney and my attorney gave their closing arguments, and then the jury retired to render its verdict.

We were sent to lunch. However, the jury said it had a decision. So we were called back. To make this short and sweet, the jury found me not guilty on all seventeen counts. The judge polled the jury to be sure, and they all agreed, so he dismissed the jury and left even before they did.

The assistant attorney turned to me, and I couldn't help but give him the finger. I felt that I owed him that for all he tried to do to me, thinking that I was just another dumb nigger. My attorney was amazed, and I told him, "That's how you win a criminal court trial." However, I knew that this fight was far from being over. Here is the docket number of my court case recorded at the Department Criminal Offenders Record Information (CORI) located in Chelsea, Massachusetts (02150) regarding all seventeen counts: # 874033 / 874049—not guilty.

CHAPTER 49

Events After My Trial

As soon as my trial was over, I first contacted the newspaper to advertise the results. After some difficulty, they did. However, it appeared next to the death section. You can now find a section of the Boston Globe that reads, "Court Officer Kenneth Walden Found Not Guilty of 17 Counts, Seeking Return of Job"—which was exactly what I did. At that time, to continue making and income, I filed for the widower's income, which you are entitled to as the receiving spouse as long as you are legally married. This allowed me to live in my apartment for a while longer. I tried everything I could to get help such as from the Massachusetts Commission of Discrimination, who took my report but was supposed to call me about my claim. Even with a copy, they said my case wasn't filed on time, and it was dismissed. I did made an appeal for it to a higher power, but had no results. You know where the power came from. All my complaints ended up the same way.

The Free Legal Defense Fund, the Legal Action Commission, Defense Lawyers Guild, and even tried the Lawyer's Guile—all said, in one way or another, they couldn't take my case. Some even admitted that I was wrongly treated, but the trial court administration had put out a denial statement. I thought even that was biased; I had to settle for no results. So I had to finally concede, leaving a bad taste in my mouth. I finally ran out of funds, which left me no choice, or at least it was the decision I came to. Even my cousin whom I had protected all her life and was now a police officer was having her problem during my trial. She called me about helping her get rid of her former boyfriend, who was also a police officer and who was still living with her and refusing to pay any rent. As usual, I responded in my way by going to the house at about 3:00 a.m., knowing that he would be asleep and having my cousin open the door. I went into his room, poured rubbing alcohol over him with a lighter in my hand, and politely asked him to leave right now and never return. My cousin was in shock, but not as much as he was. He jumped up and ran in his pajamas out the door and never returned.

My family always knew I would take care of any problem when I was called. However they didn't always agree with my methods. With me, no time for talk have. I've always been a person that never could put up with a lot of goobey-goo my way. Got to the bottom line quickly. I was always taught that as the oldest, it was your responsibility to protect your loved ones—one more thing my grandma instilled into my mind. So as usual, when it was my time to survive out there in the street, I prepared myself for just another challenge. I guess I should have noticed right then and there that my attitude about life would change.

CHAPTER 50

Shelter Life

Sometime later, I was forced to give up my apartment because all through this period, I continued to support my children; and without a job, my funds were running low. Now to begin my personal life living in a homeless shelter, this was no walk in the park. The first thing I did was observe who the head bully was. You see, I hadn't been in this predicament before. But I had many conversations with convicts that I had in my courtroom. I then had to decide whether or not and how I was going to approach this person. You see, you couldn't bring weapons into the shelter. I learned that the didn't pay any attention to bathroom supplies, such as toothbrushes and toothpaste. So what I did was take the razor blade out of a Gillette razor and put it in my toothbrush covered it with toothpaste and carried it in my bag. Toothpaste hardens, and if you put enough on, no one can see what is in it. Besides, the shelters always had someone who didn't have a lot of experience on the door back then.

Not long after being there, I met him in the bathroom room early in the morning. I came up behind him, put my toothbrush to his neck, and said, "I want you to know that I don't want to be bothered while I'm here, you got that?" He was so surprised that I think he peed on the floor. To make this short, I didn't have any problems with anyone after that. In fact, when I sat down to eat, I was given as much room that you would ever want. Remember, you are not there to make friends. You are forced to be there, but not stay there. This was a sign that I had become very negative about life and was becoming a threat to people.

I did find out that you can get a job in there if you become noticed. So you know that was just down my alley. I was hired but was tested by being given the worst job: cleaning the bathroom after it was used all day. Because of the circumstances, I couldn't complain. Thank God it didn't last long. Let me point out that's a test of whether you want to survive or not. After about a week, I was given the job of being in charge of the donations. The main rule was that no one was allowed to get anything without the head manager's approval, so you know how I listened and followed the rules exactly.

On one occasion, a junior manager came in looking for clothes for a friend. I immediately told him the rules, and he responded that it didn't apply to him. I responded that he must get an okay from the top manager who had given me the instructions. He got very upset and left. I immediately called the manager in charge and told him what had happened. He told me that I handled it correctly, and he would take care of it. So I continued doing my job and eventually got more comfortable with my situation. However, not so much so that I didn't continue to loot for better. I was enjoying better living quarters; great food, like pancakes for breakfast and steak for dinner; color TV; and free time when you weren't working. And by the way, I was getting a paycheck, not much but it was a start. However, it was 1989, and I wasn't getting any happier with my forced situation. To tell you the truth, it was starting to affect me mentally. After about six months, I started to become very angry and much more silent. It wasn't long after that if anyone came close to me and my space, I became very upset and started to respond very negatively.

I eventually punched the lights out the person that I had a run in with months earlier in the clothing area. Not that he didn't deserve it, but the rules said "no fighting," and I knew I was wrong before I hit him and knocked him out. I had to leave, but the manager helped me by getting me into the veterans shelter that was new on 17 Court Street, States Street, ironically in sight of the courthouse that I worked at. That was the cause of me being in this situation. You know how that helped me out mentally, right, seeing many of the same people as they walked when I went to work there? I can tell you that there were many times that I wanted to reach out and touch someone, if you know what I mean.

The shelter accepted me because I was an honorably discharged veteran, and I became number 150 on the list, which of course totaled now more than about ten thousand. Another strange thing that I found out was that my father used to work there as a doctor in the 1950s. I never knew that at that time. I have had a dream that all of us homeless people merged on the Washington Male and camp out like they did in the 1920s when veterans demanded their rights (appendix 7 will give you an understanding of my thoughts).

The shelter was just being put together nothing like today. But I did find work there as a building worker in South Boston at a condo building called the Foundry. I was soon given a studio apartment. Not bad for someone in my situation. I was just a couple of blocks from the D Street Police Station where my cousin worked. I began to get comfortable working there and going out to do my grocery shopping, going to the bank cashing my check, and getting a money order to send Margo for my daughters' care. I had a work van to drive all the time. I was responsible for the servicing of three buildings, the Welfare Office in Dorchester, the office for Children Services in Fields Corner, and of course the Foundry. My day started at about 5:00 a.m., which of course didn't bother me. It always seemed that I always had to get up at that time

At the Foundry, I had much responsibility: taking care of the trash bin, servicing the four floors and keeping them clean, keeping a beeper to respond to take care of any emergency, including a locked-out tenant (also any emergency at one of the other buildings including faulty alarms and any other problems). In the wintertime, I was responsible for snow removal at all the buildings and the special care of the side of the Foundry building where the "new boys on the block" practiced and maintained their office. I used to see them, but it didn't mean anything to me primarily because I wasn't benefiting from their singing. Also, I didn't directly work for them; I just took care of their office needs, being treated like a black man who was seen but not seen, which was natural during that time; and even now, I don't think things have changed much, except it has included other people today. After about a year, when I had continued to send money to support my children, I was promoted to building supervisor, but I did the same work, so it only meant something in name, as you would expect. But I was smart enough to know that I had to be thankful for what I had or at least appear to be. I only got the position because the person who was the supervisor was discharged for drug use.

I want to take this time to tell you the real story of shelter life from one who lived through it. To begin with, everyone should understand that no one wants to be homeless. Most of us are not directly at fault for it happening. Just take a look at the Trump government shutdown. How many working "secure" black and white folks felt about the fact that you were just a breath away from being homeless? For many, you were; and for some, you still haven't recovered—all of which didn't happen because you did anything wrong. This is a perfect example of how homelessness happens. It's a whole different world. In order to survive, you must put aside your proper upbringing and that feeling of being poor if you want to survive. You have to use anything or anyone you can to learn the ropes but at no time drop your guard. People will eat you alive. It's a dog-eat-dog environment. You have to learn survival. Taking care of you and yours by any and all means is your primary responsibility for your own good—learning where to go and how to get what you need like food, clothing, showers, health care; where you can lay your head and close both eyes; a clean bathroom; and if you are very lucky, where you may be able to earn an honest dollar. Yes, this is just one of the things that you must quickly learn like yesterday if you are going to begin your self-survival life in the homeless world.

All these things and more, I was fortunately able to learn mostly because I had lived an independent life for much of my life and, by nature, could adapt to changes without much difficulty. As I said earlier, I had to act a little unbalanced when you have been on a roller-coaster ride. I wanted to take it out on anyone that invaded my space. However, if you can keep control of your anger and use it to your advantage, it can become a useful weapon, which is exactly what I did. Most of the time. As they say, when in Rome, do what the Romans do, and I understood that very well. Keep yourself focused as you weave your way through your hell. Never give up or give in, and never waste time feeling sorry for yourself. This is the most important rule of all. That feeling sorry for yourself will only weaken your ability to possibly have a chance to move on. This is the trap that makes most homeless people fall into the world of hopelessness. That is the constant companion of homelessness. You must keep them apart, or you'd never have a chance to get out.

I was fortunate to have that drive because of my teaching from my grandma. As I have said from the beginning, without her guidance, I would have died or been killed long ago. God gave me special help, and as I have gotten older, I have come to my senses to realize it. I even went to Solomon Carter Medical Center, which was right next door to university hospital, the last hospital that I negotiated a contract for. As you can well understand, I would be lying if I said it didn't bother me. But I did stay clear and out of sight. My cousin Raymond Walden worked there as mental health doctor. I didn't know that until I met him inside the building one day. We were both surprised. He asked me, among other things, if he could be of some help. I told him not to worry, I could handle this. He, knowing how I thought, didn't push it but said if I ever needed his help, he would always available. I did feel secure that if everything failed, I always had an ace in the hole. Knowing and using it too soon is not the right thing to do. Remember, I used that method in my court case, so it proves that knowledge is power.

This says that life's lessons are not learned in school but through listening to the old folks and keeping your mouth shut and your eyes and ears open—just like our slave ancestors did. How do you think they survived? Lessons are all around us if you take the time to notice and question everything before believing or agreeing what you are told. This is something that's in my blood and has served me all my life, which I will show you later in this book.

Services for homeless people hasn't improved since my time, and it has become a big problem and growing faster than the services they are supposed to provide—especially when our government shows little interest. I often feel like things won't change until the homeless are living on the president's personal property, and there would be so many authority couldn't remove them. All of a sudden having to have to share a toilet with about twenty people and a public shower with even more can drive a weak-minded person crazy. Remember all these things the next time you cross the street so you don't have to come in contact. "There but for the grace of God go I"—keep that in mind. Nothing is promised. Most people are living one paycheck away from the street. The one advantage that homeless people have over everyone else is, they are at the bottom and have no place to go but up or on your property.

If you believe that you can't become homeless, just look at what happened during the December 2018 government shutdown. Many people became homeless for the first time, and the problem for many of them is that they still are. A paycheck doesn't guarantee you won't become homeless. A good degree doesn't mean you won't become homeless. Wealthy family members won't guarantee you that you won't become homeless. There are no real guarantees. Only the possibility of all of us making sure that those that we elect to protect our interests, not the interests of those who have too much, will help us to become more secure. The way we elect our government officials dictates how we are treated. So it is our responsibility for ourselves and our children to see to it that we are treated with respect.

What I have included in appendix 10 will give you information, a national report on the homeless. I hope this will scare you enough to want to do something about it.

Returning back to my job at the Foundry, at some point, I met one of the ladies employed at the Children's Family Services at Field Corner who asked me if I could put up some bookshelves in her office. I agreed and put one up immediately. It caused an internal problem. If I had known that it would cause so much trouble, I wouldn't have done it.

First, the lady that I had put the shelves up for called me to see that her shelves had been torn down from the wall. Then a woman that worked in another office wanted me to put one up for her. I thought about it for a while and then told her if I had some time, I would be happy to do it. I didn't want to get involved. When I got back to the office across the street from the Foundry, I told my supervisor what had happened, and she said she already knew, and I shouldn't have done that. I asked why not, and she just said stay away from that woman. I was puzzled by that because the lady seemed to be all right with me. After that, I noticed that the attitude of my supervisor and the big boss changed toward me. I didn't pay it much attention at first, thanking it may be that they were having a bad day because I didn't come into the office much, and so I thought it couldn't have been anything to do with me. Boy, did I have it wrong.

The next day, my supervisor beeped me to come into her office. In those days, everyone used a beeper on the job. For those who don't know what it was, it was a calling device that showed a telephone number and a code number to indicate how soon you should respond. When I arrived, she asked me to sit down, something that had never happened. She then told me that she was giving me a work schedule to follow, showing where and at what time I should be at different locations. I responded by saying that this schedule would never work because every day something would come up that take more time at any one property, and if I didn't take care of it at that time, it'd cause additional problems. She didn't want to hear that. At that point, I asked why this is happening; she said that this was the boss's desire. I knew then it had something to do with the lady that I helped. She was very attractive, and I felt that there was something going on between them. Nevertheless, I took the schedule and tried to follow it, but immediately it caused trouble, mainly because it was wintertime, and servicing the property took on a whole new light. I had to do things that caused more time than the warmer months. They were somewhat embarrassed by these events. I didn't say anything, but I did laugh to myself. It reminded me of the story my grandma told me about the white slave owners giving the broth that was left from the veggies. As usual, my grandma was always on my mind. Thank God.

It became increasingly difficult to do the job comfortably after this latest event. Now it was no longer a happy place to work. It wasn't too long that my supervisor came down with breast cancer. I felt sorry for her. However, it became even more difficult to be around anywhere, especially at my apartment, which was across the street from the Foundry. It wasn't long that I found out that an elderly woman that had lived there for years and was a friend of my supervisor started spying on me. I discovered this after I had to go out about 2:00 a.m. because the fire department called me because the alarm had gone off at the Welfare Office in Dorchester, so I had to go to turn it off. The next morning, the boss called me in and immediately asked me where I was at about two thirty this morning. I told him what had happened, and he said by mistake that his friend had told him that I had left. Her third-floor apartment looked over the parking lot where all the cars were parked, so with that, it became clear that trouble wasn't far behind once again. My past was once more on the horizon. As you can be aware, I started to begin to watch my back because this was South Boston; and even though my cousin worked a block away, even she didn't have it easy. During those days, it was normal for all state jobs to be always filled by whites, at least the better ones. My current problems was a testament to that.

In early 1992, I was confronted with a new problem. The office informed me that they had hired a new building supervisor, but I would be sill working and keep my apartment. Plus, I had to train this white relative of the boss. Of course, I was very angry but knew I had no immediate choice. It wasn't long when he made a serious error when he was trying to get more heat from the Welfare office boiler. When the office found out, he blamed me for the problem. I said, "I have serviced that boiler for about two years with no problems. Do you think I forgot how to do it?" The response I got from the boss was typical of what a white person says when they are wrong. "Let's not worry about it," like I had to worry in the first place. You know after that I wanted to punch him out but didn't because, although it may have brought me immediate Satisfaction, it would put me in jail without a trial. But he knew he had to stay away from me. I had developed an angry look on my face by this time in my life in part because of my current conditions, so I knew I had a quick temper. I was asked by the boss to plant some flowers for the very lady that he had said that I should stay away from. I knew right away that they had an on-the-side relationship. What follows changed my life forever.

When I went over to this lady's house on, I believe, St. Paul's Street, Brookline, she had a basement apartment with a small area in the backyard where you could plant flowers. I arrived about 5:00 p.m. It was in the summer, so there was plenty of sunlight. There were about three crates of flowers. I spent the evening planting while she talked about general subjects. After I had had finished, she asked me to come in and have some supper to pay me for my help. I didn't see any harm in it because I would enjoy someone else's cooking for a change. After the dinner, I was sitting in her living room. As I remember, The Witches of Eastwick was on, and her doorbell rang. She said to stay where I was; there was no problem. Before I knew it, I was hit over the head with the butt of a gun a few times, which knocked me to the ground. The next thing I remember, the police were there looking after me. I was rushed to Beth Israel Hospital, where I had several stitches and my two front teeth pulled; plus, my eyesight was off focus. When she came into the hospital room, she started apologizing and said it was a jealous former friend—as if that made me feel better. After, the police took my statement. This moment stands out because it was just before my birthday.

I discovered that the man was a black court officer that I knew who worked in the civil court where my wife had the ticket problem. Can you beat that? It also told me that this woman was a player. When I returned to the Foundry, I couldn't see clearly and had a serious headache. That was just the beginning of my trouble. Since I couldn't work, and I knew that the boss wanted to get rid of his side-affair problem, I was let go. So I had to return to the veterans homeless shelter. There was no explanation given by my former employer, and no one wanted to hear what I had to say.

I immediately filed for my unemployment. At the first hearing, I was denied. At the appeal hearing, I was denied, so I read the fine lines of the appeal hearing documents and discovered that you can appeal to a third hearing with the board. So you know me. In that hearing, I produced a letter from my doctor stating my condition and the accident that had happened, including the time and place; plus, I had the surprising help of one of the board members who recognize me from my days as head of my labor union. Needless to say, I won my case, and the board member said, "I see you haven't changed a bit." I knew what he meant. I was able to collect retroactively back a month and a half. So you know how I felt. Remember. I was still thinking about the welfare of my daughters.

The criminal part of his trial was held at Dedham District Court, where he pleaded guilty and was sentenced to five years' probation and restitution—then in Boston Civil Court, where his wife, who was a lawyer, represented him.

This was a very difficult place for me to come and testify. Remember, I had worked in that courthouse, and a court officer who was there remembered me. He did help telling the judge about my past problems, so the judge said that, because I appeared upset, which he could understand, I could remain seated in the chair I was in. I have to say, all my bad memories came back. I was surprised that they allowed this judge to hear my case. My lawyer or I said nothing. In short, he was found guilty and was ordered to pay me $13,000 in monthly payments for hospital expenses and $56,000 for pain and suffering plus he was placed in five years probation. However, my lawyer told me that the final judgement was $55,000. His name is Peter Olmstead, case number C.A. No. 93-4108-G. The $66,681 I never received. I didn't know this figure until I recently picked up these records.

My attorney said he couldn't be found. I have always found that very curious because he was on a five-year probation and was paying me monthly $100 for medication and his wife is listed as his attorney Lenda Ormstead notice on the court records. So I could never understand why she couldn't find him. However, I wasn't myself during that time so I didn't pursue it as I should. But now I have included the documents from the court that says what my judgement is and that an execution issued 2/12/1996, page two in the Appendices I am not saying that my lawyer did anything wrong but these are the questions that I feel are important. I have reasonably reached out to my attorney for a statement concerning this issue but as yet haven't received a reply. I even called the Bar of Overseas and asked them some questions. Such as how could I find out if the money was paid. They said the case was too old to check. I responded by saying that there should still be a record. I asked couldn't they inquire my attorney to present something to show the results. They said there may be no records available. They finally said that I should forget it and they couldn't help me sounded familiar. This occurred in September of this year (2019).

This is a perfect example of what can happen to you when you are Black and thought to be stupid. A form of racism is always around.

CHAPTER 51

Pain and Suffering

Pain and suffering, which I never received because he didn't have a home address where you could serve papers when it became time for him to start paying that part of the money, and the law said this was what I was told; you couldn't use tax returns to find him. I never believed that, but everywhere I asked, I got the same answer. But I did start receiving my monthly money plus because I had begun receiving mental health treatment. I went to the state house building and filed for victim of violent crimes. After a couple of months, I received a check for $2,200 for my loss of income. I sent this money home to my folks to hold. After that, I went to the federal building at North Station to file for SSDI, and as soon as I went to the desk and said my name, the person at the desk said, "Hold on a couple of minutes." Two ladies came from upstairs and asked me to join them. I didn't know what was happening, but I went. Lo and behold, they said they would take care of everything, and all I had to do was just sign a few papers. I asked if I would have to appeal or anything. They said no, I was all set. But they did ask if I had been working since I was born because I was going to receive the highest amount of benefits offered.

I said yes, very young, and then I left. It had always appeared that I had always came back from living in hell with money available. I guess this was one more thing my grandma said: I would always find a way to make a penny. Boy, she had always been right. Remember, I didn't get all these financial benefits all at the same time. But it sure felt like it.

All these things happened as soon as I had begun having much more serious problems with my anger. Beth Israel Hospital started seeing me for mental health therapy that I was still going for PTSD, attention deficit disorder, and anxiety disorder. It caused me to stay away from people for the most part. I recall being home and taking it out on my brother when he asked me about my shelter experience. I started yelling at him and cursing. I couldn't control myself until our mother came into the room and calmed me down. I think I scared my brother. I did apologize but remember it up to this day. I am still upset about it, but we haven't talked about it since. I guess it shows what real brothers are.

When I returned home, I talked to my doctor about it, and it showed how serious my illness had become. He said that at least you are aware, and when you feel this was going to happen, come right into our safe haven clinic because two wrongs don't make things right. For some time, I found it difficult to be around people in general, although I was still sending money off to my children. Now with both of them in college, my oldest kept saying, "Oh, Dad, please send some money. This food is terrible." So you know I did enough for both because they were both going to the University of Iowa. However, it wasn't long before my baby girl, a name I have always called her, called and said the same thing. I asked, "Doesn't your sister give you half of what I send?" She said no. So I started sending it separately. When my $2,200 came in 1993, my mother was able to pay for each of my daughter's college bills as you can see it was much cheaper going to college in the 1990s. My youngest daughter, Leah, transferred to Bowling Green in Ohio, graduating in 1995. My oldest daughter, Demeta, transferred to Southern Illinois. She went to college but didn't graduate I never knew why, but the main thing was, I never knew she started taking classes at a community college. That created a big problem a little later. I just want her to know that if she had told me, I would have seen to it that she was supported. As I recall, my mother said that she wasn't going to school any longer and that she had a boyfriend.

At this time, it was very difficult for me because I was going through a lot of mental pain.

I did finally go home. My brother Robert was glad to see me in part because he loved to think he could live my so-called adventures, something that he still enjoyed even when he didn't want to admit it. But he's one of two people important in my life that have pushed me to write my story; he was the one that named me Jeremiah Johnson. It has stuck to this day. I guess I was the only member of the Walden family that had never had an ordinary moment. About this point, I haven't ever been able to convince my mother about anything pertaining to me, including the events surrounding my divorce. For some reason that I had never understood, I had gotten used to it, but I tried; and even with some written proof—for example, the cancelled checks that I mailed her several years ago concerning Margo—she still said that I hadn't been supporting my children. She had always found a reason to either not believe it or turn to another subject. Just wanted to let you know that it had never been easy being me.

While visiting, I got a call telling me that they had found an apartment for me in Cambridge at the Fresh Pond Apartments, so I had to get right back to Boston. I started living here in late 1993 and had lived here ever since.

I did go home in 1995 for my daughter Leah's graduation. That was at Bowling Green. The thing I remember most outside of graduating was her mother, grandmother, and aunt with her three children being there eating and when it was time to pay the bill. They sat there making no attempt to offer any assistance. Plus, they didn't give Leah a gift for her graduation. My patient and I did. I gave her a few hundred dollars. That, she said, came on time so she could get her things out of storage. I had already paid for her being at Bowling Green and the money she needed in order to stand in line and graduate.

I need to point out that this was the last time I heard or saw my daughters until 2017. I will get into it later

CHAPTER 52

Going to College

I spent my time in my then new apartment, staying away from everyone. Even today, I only know my neighbors enough to say hello. I didn't go anywhere social or have anything to do with anyone expect going to Beth Israel Hospital for therapy. Then every week, I stayed separate from any contact. I even had to admit myself twice for feeling like I was about to go off on someone. You have no idea what can trigger you off unless you have suffered from this disorder. I was being heavily medicated, but it didn't always do the job. Thank God I had enough control to know when I was in trouble. All during this time, my father had become a well-known psychiatrist in Toledo.

But by 1996, I was pushed into public life by my therapist to go back to school. I took that chance and started with the Veterans Outward Bound Program (appendix will show you my diploma). It was a beginner's program for veterans to begin the struggle to create the mind-set to go into college. I went for the interior year, plus as an incentive we received a stipend; that always helps. I received help in writing and computer understanding; for most of us, we were used to the old Underwood typewriter. At least I was. I had and still do have a problem of hitting the keys lightly. I always get three of every letter I hit, even now. We also got help in general math. By now, I have found that I have lost all knowledge of the things that I could once do in my head.

After graduating from the training program (appendix 11 will show my graduation certificate and other material associated with my attending college), I was accepted at the college. Math was my most difficult problem in college. I could figure out any financial-related problems, which was amazing since I had cut my teeth on dealing with all kinds of math-related issues when I negotiated union contracts. In therapy, I talked at length about it, and it was determined that it had a lot to do with my past. My mind was recalling my painful past troubles. I asked what was the answer, and my therapist said, "In time, you will be able to confront this problem without feeling angry and very upset." I couldn't understand. But that was part of the healing. The more I talked about all my anger problems, it didn't make it any easier to think sometimes because every year, your doctor changes, and the new one had to ask questions about what happened. This opens up Pandora box all over again. The one thing that still happened, however, after more than twenty years. You get ready for it, and it doesn't affect you as much.

As time went on, I got a lot better at math and very good at other courses. I took arbitration, meditation, and the negotiation course, for reasons you can understand; also writing, government, and human resources were on the agenda also. My most serious problem came trying to use this damn new thing called a computer. It really scared the hell out of me. I was all numb looking at it for a while in the student's study room with all these kids around me. I didn't know how to turn the damn thing on. So eventually I swallowed my pride and got some help from a student that was assigned to the room. As time went on, we got together at a time when other students weren't there. He started showing me the basics like getting on Widows and learning the most important tool: spell-check. Also how to change where you want sentences, I think they call this copy and paste. I still have a problem: printing and how you saved what you had done on your disk, something I clearly didn't know why I had it. The one thing I still can do is use more than one finger on each hand. He would laugh. He asked me how and why I could only do it this way. I told him that's because I was used to the Underwood typewriter, where you only have to use a finger from each hand, and the damn keys weren't as sensitive as these machines. But lo and behold, I did at least get used to it, except for e-mail. I didn't know how the use that damn part until this last couple of years.

My brother would e-mail me and then call me a couple of days later, asking me why I hadn't answered his e-mail, and I would say, "I never got your call." I didn't really know what he was talking about. He gave up with a laugh, saying that I would never leave the nineteenth century. I was given an e-mail address but never had a use for it. I did type out all of my paperwork using my two fingers, where all the younger students would ask how I could do it that way. I would say, "Look at the results of my work." I spent a lot of time with my advisor, someone whose purpose I didn't understand at first. A time came when she became very important; she just happened to be the assistant director in the college. Her name was Sarah Barrtlet. I remember her like it was yesterday. If it wasn't for her taking so much time and patience with me, more patience than anything else, I wouldn't have made it through graduation.

I was voted by the students to represent them on the university advisory board, being able to participate in such things as financial, tenure, and overseas trips. I was amazed to see that some of the board members knew about me when

I was union president. I even was the speaker for those students coming to college. I recall, among other things, saying how safe this place was for the women with our state police force being here. I don't even know why I even said that. But later, a state police officer came up to me and shook my hand and said, "Thank you for thinking about us." I was a speaker at some of the retreats to other colleges around Massachusetts like Boston University, Tuft, Simmons, and a few others. The students were very interested in the 1960s period of racial division and my attempt to organize health-care workers.

In my legal law class, I was reintroduced to the lawyer that represented me in juvenile court when I was fourteen. That's when my father tried to get me to give up my apartment. But the judge said that as long as I could support myself, his hands were tied, thank God that the Judge didn't remember me from my car-borrowing days. His name was Herman Hemingway. I had no problem passing that course because he asked me to fill in for him a few times when he couldn't be there. Because in some way, he knew I had been a Suffolk superior court officer. I even met Scotty Rose's grandson because I was talking about my years as a court officer in class.

Believe me or not, even Ms. Spencer was there teaching nursing. It was like old home week or all the robins coming home to roost.

After that year, I had about thirty companies where I only needed nineteen more to graduate. Companies was the same as credit.

In 1998, I was visited by a lady named Laura Booth. She was on the board of a Cambridge Community Services Agency called Cambridge Economic Opportunity Commission, better known as CEOC, located at 1 Inmand Street, Cambridge, Massachusetts. I was taken by surprise. Thank God, it was one of my good days. She offered me an opportunity to join the board. What I didn't know at that time was that she was also attending classes at the university. Like a smart, intelligent lady, she observed my actions at the college. I said I would think about it because I was told by my therapist that it can be harmful to take on too much. After talking about it with him, I decided to take it on. To my amazement, I was quickly voted president of the board. I have to say it was a very positive experience. We handled housing problems, had a meeting with the Cambridge mayor and city council. I went to the state house to speak about housing problems with our state representatives and other duties. When I was busy at the college and this board, I didn't feel that anger that had always laid under my skin. It even stopped me, for the most part, not to think about why I haven't heard from my daughters. My work at school was going well. I even got work study with a paycheck, working in the same big office as my advisor. Why is it that, for some reason, I am always able to find, even when I don't expect it, a way to make a dollar? I guess I shouldn't complain.

Now comes the sad part. I was called to come into the hospital for my yearly physical, something I hated because of that glove treatment. I know a lot of you feel the same way. My nurse practitioner had gotten to know me very well since I'd been going there. So it was like an order. On my birthday, I went and had the full treatment. She called me soon after and asked me to come in. I felt this was unusual. But I came. She said I had some good news and bad news. I said, "Give me the bad news first." She said that I had prostate cancer. I almost fainted. Then I said, "What can be the good news?" She said, "We have discovered it in time. So with an immediate operation, you should be fine." I guess looking back, I should be thankful that it was discovered so soon.

You know if you ever had this experience how destroyed you feel. I had to make an immediate decision. I called home spoke to my mother, I guess looking for some sympathy, but I should have known that wasn't going to happen. She just said you know what you have to do just like you have handled everything else. So a month later, May 15, 2000, I had surgery. However, I went to see Sarah at the college to say goodbye because I thought that my college days were over. But she tried to convince me more than once because you know I am hard-headed that I will be able to return and have one-on-one study with her because she was my advisor. I heard her, but I wasn't believing it at that time. Needless to say, all my mental disorders came forward at the same time. There I was feeling like my world had come to an end, being that I had gone through this so many times. I hadn't heard from my daughters, no one in my family here, and I was feeling sorry for myself, something that was very new for me. I had a rough time dealing with it and was very uncomfortable. But as time passed and with the important help of the Cambridge Board, I gradually came back to life as best I could and went back to college. However, there were serious health-related difficulties. It not easy for me as a man talking about it, but the truth had to be told. I immediately started have urinary problems, started to live in the bathroom. When I left the house, I had to make arrangements to use every bathroom going and coming home. I wasn't told about the dependents. If I had, I probably wouldn't have used them as I was too proud to wear one. When I went to school, I had to make arrangements for a state police car to be at the train station to allow me to go in the bushes without being seen as a pervert. They thought this was a funny request but said I was using my head in calling for their assistance.

As time went on, I got a little bit in control. However, not completely, so you know what I had to do, right?

Unfortunately, because I was spending so much time adjusting at school, I had to finally give up my position on the Cambridge Board on a regular basis, but they still wanted me to attend whenever I could. To tell you the truth, looking back, I believe if I didn't have these people in my life at that particular time, I don't believe I would be here now.

I graduated 2004 with a BA degree in labor and human resources. I drove Sarah crazy before graduation, asking her if I had done everything in order to graduate. She kept telling me yes and relax. The problem was, my brother had planted in my head that sometimes they just gave you the folder if all your paperwork and bills weren't paid. So you know that disturbed me. My mother and father attended my graduation. I know my father was especially pleased because that made me the last one in my family to graduate (appendix 12 will show my diploma).

It was my intention to continue on to get my graduate degree. However, my professor told me that I would be wasting my time because all opening would go to much younger students. I was very upset about this. But it was true that during this time, there was a hiring shortage. Many people had to take low-wage jobs who were highly trained for much better positions. I guess it was another case of wrong time and place. So I had to deal with it. But for a long time afterward, my brother kept pestering me about different places for me to try, not listening to me when I would constantly tell him that I had tried everywhere I could think of to find employment in my field. The funny thing was that he had constant trouble finding permanent employment in his field. Well, I guess it's a case of "do what I say, not do what I do." I know his heart was in the right place, but it did almost drive me crazy.

CHAPTER 53

Concerning My Daughters

The last time I spoke to my daughters was when my youngest graduated. To this day, I don't know the reasons why. I had nothing but daily pain, wondering about them. I had told myself many times that because they are so much like me they are out in the world finding their own way. It sometimes had worked for a moment. Even now, it's still a major topic of discussion at my mental therapy appointment. There was the time sometime in the middle of 2009 as I recall, I had gone home to visit, and my mother pulled out a picture of my youngest, married with my granddaughter, grandson, and her husband, posing for a picture in our den. My mother may have thought she was doing the right thing, but for me, it was like a dagger in my heart. She couldn't or wouldn't understand how left out I felt not being told by my daughter or about this and how my mother so calmly was showing me this. Only my brother knew how I felt. My mother once again didn't think that much about it, and my father took his usual position of saying nothing. So knowing that I couldn't change anything, I dropped the subject for the rest of my usual five-day visit. But when I got home, this was one of those things that hurt me so badly that I had to go into the hospital and stay a week so I could get full control over my emotions.

From that time until today, I have suffered over the facts of my daughters' absence in my life, always feeling guilty for something that I might have done.

I began taking my yearly trip home, having my brother always pick me up whenever the train would finally arrive and take me back to the train station. That's the way of train travel. Going home every year for every trip thereafter, I had developed a fear of flying.

Thanksgiving or Christmas and New Year up to the present, going home always has a special meaning for me, such as my father passing away in 2012. Thank God I was there to see him just before. I whispered in his ear that he was my hero. He smiled, and my mother said for the next two weeks, he was feeling pretty good (appendix 14 will give a history of his work). If I remember correctly, it was after that I started sending my mother flowers every Valentine's Day, Mother's Day, and birthday up to today.

The very next year, my cousin Karen passed away. All these deaths seemed to happen right after one or another. I think of these things at least once a day.

Then with the convincing of my brother, of course, I went to see my mother, who was dying, and made my peace. There was the Christmas and New Year's that I went home, and all my sister and brothers asked me to go with them to see the fireworks. For some reason, I decided to stay home with my mother. Thank God, I did. My mother was going up the stairs and fell back down. I immediately ran to her. She was lying on the floor with a big bride on her head. I tried to help her up, but as usual, she was being stubborn. She said, "I'm okay." I called my brother and told him to come home while my mother was fussing with me. I decided not to listen. I felt that she was just being her stubborn self, and as long as I had her fussing at me, she would stay focused. My family came home. My brother said it wasn't the first time. However, I said it was the first time for me, and I couldn't understand why she was so resentful. He said, "You know that's her way." They took her to the hospital where she stayed a couple of weeks because of a broken wrist and ribs. When I went to see her at the hospital, she still showed me resentment. I can't remember her ever saying I was right or even thankful. I took it as a sign that was my cross to bear.

All this time, my brother was trying to convince me to use my usual ability and try and contact my daughters. It took until early 2017 for me to build up the courage to do it. This is one of many examples of our closeness. Neither of us would ever admit it. I had gotten somewhat used to this new cell phone that he forced on me. I learned about Google search. However, he said that it would take time to type in my request. I wasn't going to accept this. I learned that you could just talk to it, so I did, much to his amazement.

I did find their addresses. However, it said that their father was Kenneth Davis. After pondering for a week, I called the numbers provided. I did get thrown to my baby girl and called, holding my breath. I left a nervous message saying, "If this is my baby girl, this is your father trying to reach you. So please give me a call." My youngest called back, but I was too scared to speak. Finally, we spoke and began to relate. She started calling me every Sunday morning, giving me a chance to talk to my grandchildren, and even sent me a birthday present that told me she hadn't forgotten me. She started telling me that she was looking for a new car because her car that she had for twelve years was giving her trouble. So you know what a loving dad does. I said I would help with payments. She said, "No, Dad," but you know who won that conversation. So I did help right up to that November when we arranged to meet at my sister's house for Thanksgiving.

CHAPTER 54

The Gathering

So after that call, I received a birthday card from both of them along with pictures of the lost years of the growing up of my grandchildren. I thought that was a great birthday present. But my oldest had no return address. But because of Google search, I discovered exactly where she lives in Chandler, Arizona. I even sent a birthday card to that address that October, but it was returned. I still have it. However, I now have one back along with my grandchildren, so I guess a half a loaf is better than none at all.

As it got close to Thanksgiving, I purchased my ticket as always getting my senior citizen discount ticket, the serviceperson asked me about the points you receive for frequent travel. I said I didn't know anything about it. When she explained, I asked her could I get retroactively because I had been traveling this route for years? She laughed and said, I wish should, but I couldn't, so I just accepted my last two trips as help with my purchase. When I arrived home and told my brother and mother, my brother said I always found a way to get more for less. Of course, my mother took his side as usual, something that has always been the case.

We left from Toledo for my sister's house in Cleveland, with me nervous all the way with my mother, Robert, and his daughter, whom I love like she is mine, except she has gotten too big for her britches, joining in on all the yearly teasing I receive. Upon arrival, I thought they had beat me there.

When my daughter and granddaughter arrived, all my past hurt not seeing her quickly disappeared. It seemed like it didn't matter. It was now that was important. We kissed, hugged, and cried a lot. Finally, we let each other go, and I turned my attention to my beautiful granddaughter. She acted like she knew me; it was very surprising. We couldn't stop holding hands, or maybe it was me. What's the difference? After we settle down, my granddaughter when to a soccer game with her new cousins. My daughter and I spent time getting better acquainted. I am being careful not to tread on bad soil—meaning not bring up anything that may spoil this visit. I quickly learned that even with the long absence, she was just the same baby girl that I knew, still using tomato ketchup on her eyes, had to have that one cup of coffee in the morning, and I could tell that she was as independent-minded like me. This was all I really needed to know that she was and is my loving daughter. Nothing else really mattered enough to bring up at that time.

When my granddaughter returned, my brother told me that she asked him why hasn't she been told about us. He said to ask me. So when we could get away, we went into the den and talked. I told her that many things had happened that are now not important if you really love and miss someone. She said she understood. I also said that when you feel it is the right time, it would be best that she ask her mother and grandmother since she didn't have a chance to see me for herself. Even though I wanted to say a lot more, I quickly noticed that she had a lot of me in her—not waiting, wanting to know about something right away, not being afraid to ask. I could tell that she was being raised the same way I had taught her mother. I did my best not to say anything negative about her grandmother, knowing it would open up Pandora's box. I think she knew that. She asked about her aunt Demeta, my oldest daughter, why she had never spoken of me. I was kind of surprised that she was that observant. I told her it was a hard problem that I never understood myself. I said someday her aunt would get past any difficulty that she was having. She said okay, so we left it there to continue to have a great visit.

We took a lot of pictures; she even showed me how to take a selfie. She spent all her time laughing, getting used to all of us, teasing me as usual. She even, by suppertime, join in herself. I asked her to help me, but she said no because she was learning a lot, and she had never had that much fun. Now who does that remind you of? Karen. She sat next to me at supper along with my daughter. My mother and all the rest of the family were very happy about this reunion. My brother, of course, had the most to say. I have to admit that he has been my biggest supporter. Even though I never tell him, because he has always been my student, not my teacher. You know what I mean.

We spent a few days together before everyone had to get home. Going back to Toledo, I guess I was too quiet for my mom and brother because they teased me all the way, trying to keep me thinking about the happy moments I just had. I understood why they were doing it. I guess it worked to a point. I had to talk to them about all the other questions I have but didn't bring up. They both said it was a smart move because I had to understand that it was rough on her also. This made a lot of sense, so I stopped talking about it and enjoyed the two-hour trip back. I didn't get a chance to meet my grandson because my daughter had told me that he might have to work. When they got home, my daughter said that she would call me as soon as I returned home.

As soon as I got home, I was anxious to receive her call when she did we arranged that she would continue to call me on Sunday mornings after she did her morning jog. You see, she was running ten-mile meets and had won a couple out there in Chesterton, Indiana. She sent me pictures. I took this opportunity to ask if I could send my granddaughter a monthly allowance. She agreed. You noticed I asked because I knew that being my daughter, she had strict rules, mainly because she had been a single parent for some years. I believe my mother had told me sometime earlier. I didn't bring it up during our get-together. This was a new beginning. Earlier that same year, I was communicating with a very important young lady on, believe me or not, Facebook. I will tell you all about it soon. That Christmas, I received a beautiful gift that I had to open right away. I never had patience when it came to about everything. My family has always teased me about it. Now my daughter and granddaughter began calling me. During my Sunday conversations, I was able to begin a relationship with my grandson. I could tell he was confused or at least distant, which I felt was natural.

There was one question that continues to disturb me. Why was it that my ex-wife had tried to distance my daughters from me only after my baby girls' graduation? You see, because one Christmas, I believe after her graduation, I was home, and she called my mother as they always did something else I never knew. She said that she and Demeta were coming to visit. When my mother said I was there, she immediately called back and said she had caught a cold and didn't think she could come My brother got right into the real reasons.

My mother got a little upset with both of us being so negative. She couldn't understand why I was so upset. So my brother soon let it drop My father didn't say anything that I can remember. He would always say I have to live here. Looking back, I guess for him, it was the right thing to do.

However, that visit did leave a dark cloud. As usual, it was a major discussion at my next therapy session. This wasn't my most important problem. More importantly, I remembered clearly that as far back as my children living in Boston up throughout their college days, they were calling me for financial help at school and earlier, calling me dad. They might have known what was written out and shown on Google about their father, Kenneth Davis, plus his relatives—with no mention of me. Even my ex-wife looked me straight in the eye while attending my daughters' graduation, who knew and most likely did this, knowing this was an insult, to say the least. My brother has pointed out that it may have been a result of the state having only that information in their records. He may be correct, but it doesn't convince me. I have to hear it from my ex-wife for myself.

Something that I don't think I will ever forgive, if she had anything to do with it. As you can see, I have never been able to stop needing to know, as much as I have tried to even now put it behind me. I hope if or when I find out, there will be an answer that will cause me to ease my negative thoughts. But right now, I have tried to focus my attention on being very happy with the way things are.

All through 2018, I enjoyed the growing relationship that I have with my daughter and of course my beautiful granddaughter. However, in therapy, I had trouble keeping my excitement in check. During that time, I got so used to my daughter's call on Sunday that I held my phone in my hand, waiting—not embarrassed to say it. Finally, we arranged for me to get a call every other week. I didn't mind too much because by then I was messaging my granddaughter all the time. See what it takes to make you learn. I had already learned Facebook, Messenger, and Google. So you see, I had joined the twenty-first century, at least in some things.

By the summer, we had arranged to meet at my sister's house again for my niece's college graduation from Ohio State. My grandson and my daughter's male friend came. Oh, what a time that was.

I met my grandson for the first time. I could immediately see that he wasn't comfortable, so I didn't force it, in part because my daughter had been married to a white man, and I had thought that it could have been part of his distance. But my granddaughter didn't seem to let it bother her. I did consider that he was nineteen and had just started working. I even thought of the school difficulties that he may have faced. Most important, he never knew that I had ever existed. Remember, their mother had never mentioned me.

All these things I considered. He did finally relax to some degree but was amazed how our family teased me mainly, something that my daughter and granddaughter had already experience and were looking forward to it.

Now for the real fun, my daughter introduced her friend Lee to me, and immediately I began inquiring about him, his life, and his relationship with my baby girl—no if, ands, or buts, just like a possessive father. He wasn't too surprised. After all, he was a grounded man, as I found out immediately. He had grown daughters of his own, something that worked in his favor. My family tried to get my daughter to rescue him. But she, laughing, said, "Oh no, I expect my father to take him over the coals." I warned him before we came to be prepared for it. Finally, I did let him off the witness stand. He wasn't upset and then joined into the fun. Even took some shots at me, which told me that he was for real. All of this my grandson saw for the first time. I knew it would take time for him to digest it all. But my granddaughter was enjoying every minute of all of it. There were over twenty of us in the immediate family. We took some group pictures that even Lee was in. This was the way you passed

the test. This August, we are celebrating our mother's one hundred years old birthday. This will be an event to remember. There is a picture of her and us in the appendix, along with her car, which she still drives.

By this time, I felt much closer to my family and tried my best to not constantly dwell on the painful problems that's embedded in my soul. Right now it's painful writing about it. Part of my reasoning is to help others understand that they are not the only one that carries deep wounds that stay just under the surface. Unless you have been there, you will never know how it affects your entire life. So we should stop passing judgment on people when we don't know or haven't suffered from this kind of difficulty.

Now having said what is the most painful part of my having part of my family back comes the reality of the suffering that goes along with it. Like they say everything comes with a price, and no bad deed goes unpunished, or no pain no gain So I have to take a break at this moment to stabilize myself, just keeping it real. I hope that my family if or when they read this have an open mind because they aren't the victim. As I taught, you face your problems with an open mind. Your dad needs your love and understanding also.

I want to send a special message out to my daughter Demeta. I have thought of you every day since we last saw each other more than twenty years ago. I am still confused over what has caused you to want to not communicate with me, but I can assure you that whatever it is, it wasn't done purposely. I can understand your feelings, whatever they may be. As you can see, I went through a similar experience. It took me many years to come to see that I was somewhat young and misunderstood the events. At least how I looked at them during that time. You did send me a birthday card in 2017. I sent you one to Chandler, Arizona, address in October the same year, but it was returned. I wish with all my heart that

you will forgive anything that you think I have done and return soon. I remember all the great times we shared. I hope you can recall them also. I love you very much. Please return.

That November, my family returned to my sister's house for Thanksgiving. My grandson was much more involved generally—laughing, teasing me, or joining in on other conversations, talking about his interest and in general just being a part of the family. Of course, my granddaughter was just herself acting like she had been member for all of her life. Evan Lee had no problems showing his feelings. My daughter was showing that she was in charge of her family. My brother took over as the big teaser, including Lee in some of his comments. However, Lee returned the favor at the dinner table after we ate. Everyone noticed and committed on it jokingly. In general, everyone had a great time. By the way, my daughters' birthday are on October 13, 1968, and November 8, 1972, my grandson's is on November 7, 1998, and my granddaughter's is on November 30, 2002—a very expensive mouth.

My artist friend saw to it that they all received personal birthday gifts. My daughter, granddaughter, sisters received necklaces, and my mother received a personal flower vase. That is shown in Karmai Alexander's artwork (appendix 16). My grandson received a hand painting of Tupac, something that his sister said he would really like. I have treasured her assistance in this area, and her general help kept me focus on the positive side of life for the last three years. I will say more about her later.

CHAPTER 55

Pension Fund

First, I want to say that I had long forgotten about my union pension. In 2016, I wouldn't have found out if it wasn't for a new therapist going through my every two interview. We happened to be talking about my involvement in the founding. So I decided to look it up on Google. Lo and behold, there was nothing about our starting the union, and it said no information about the1970s. So you know that piqued my interests. After much checking, I discovered that SEIU had taken over Massachusetts in 1989 with the agreement of Henry Nicoles. Plus, the New York districts, including Doris Turner's, had been transferred, and she had retired. In the history of our union, it mentions Leon Davis and Henry Nicoles, but no one else. In that section, it goes on to talk about the growth of the union but doesn't mention anything about Massachusetts. It also said no records concerning the 1970s. What it says about Massachusetts is that it started with SEIU and that Tyrek D. Lee was the first black president. He wasn't even born when I negotiated the contract at his hospital. Speaking of him, there is an appendix 8 showing his 2017 problems while representing our union.

You know that's wrong. I began to investigate further. I found material that said in the late 1980s, there were some major changes that had been in the works for several years, something that I had my suspicions about. My therapist asked, did I ever get my pension? Well, that's when it all started.

I contacted the lawyer that handled my assault case against Peter Ormstead, and she agreed to take my case. To make this story shorter, she sent several letters to the National Union headquarters in New York to the National Union headquarters to Henry Grissom more than three times before he directed our concerns to the Massachusetts District. After repeatedly trying to get a response, they said they had no information. We submitted certified letter to Henry Nicoles in Philadelphia District 1199 C. No response until three months later. Their attorney said that they had no information on this matter. This material would provide a copy of both exchanges, even with these documents showing my involvement. They refused to acknowledge my existence. This went on for more than two years. Appendix will provide written documents, showing the many letters sent.

This past year, my attorney found out that UMASS Boston has a department that handles these matters, just so happened to be where I had graduated. So after contacting them, they took on my case at first they ran into the same responses. But with the help of Cornell University, which keeps records on everything, they were able to pull up things like my picture at a 1975 AFL/ CIO executive board meeting in New York, my employment records with the union, and a Boston Globe 1973 interview when I was secretary-treasurer of the union. These things will all be part of proof in documents beginning with appendix 10.

Lately, I had been informed that Henry Nicoles had now admitted to knowing me, but we are still at odds over my pension. They are saying there are no records for the period of my employment. The chief lawyer has recently said that they are taking my case into federal court—once again a battle to be fought and possibly won.

I have no idea if or when I may get my pension, but since it has been more than thirty years, I guess I can wait. Once again, it shows that you should never throw away old papers. You never know when they may become important, something that us senior citizens are well aware of. I seriously believe that they thought I had died and gone to hell. Or at least where I couldn't come back to haunt them. I have just received a denial letter from the NY office saying that I didn't qualify for a pension.

CHAPTER 56

My Brother Robert

I have saved the best moments of our long relationship until now. I think it deserves a chapter of its own. To begin with, we have always been known as Frick and Frack, or Nip and Tuck.

First, I want to remind everyone that I have suffered with post-traumatic stress disorder, attention deficit disorder, and anxiety disorder from my attack in 1990s. I say this to remind everyone how impossible it was for me to get employment, something that my brother never understood at that time. Also there was a hiring freeze. All affected me from doing what my brother kept hounding me about. I believe it was because he was young and couldn't see the forest for the trees. Now I believe because of his own working problems, he understands and never wants to talk about it. Age brings knowledge.

There is always something new happening when I come in by train, going or coming. For example, the train is never on time. For over twenty years, he has picked me up at around 7:00 a.m. to about 11:00 a.m. and taken me to the station to return home at 3:00 a.m. until noon, depending on the weather.

Every trip has its own funny story. For example, one trip not too long ago, we were at the station waiting for the train for Boston around 3:00 a.m. A lady that weighed every bit of 450 pounds was there sitting and leaning on my luggage. My brother as usual said, "There's your honey waiting for you," from clear across the sitting room that echoes when you talk late at night. Everyone sitting there heard him and looked dead at me. The lady smiled and motioned me to come and sit. I didn't know what to do. So I went outside on the platform and waited for the train. He came outside saying, "You still have to get your luggage." When the train came, I got my luggage and rushed to get on. My brother stayed on the platform to see how this poor lady was going to get on the train when she couldn't fit through the doors. He later told me that they used a luggage lift and put her in the baggage car, and the train leaned to one side. I don't know how true that was, but I do know that he went home and told our mother all about it. That was and still is a topic of discussion.

Another memorable event was during the holiday travel season. There was always more ticket sales than seats. So I always called the ticket checker before boarding if my reserved sitting is available. The over-the-phone person and the ticket checker said yes, so I boarded the train expecting to be seated in that seat. Well, that didn't happen. There were students setting and lying on the floor in the seat reserved for me. I asked the on board attendant what I was supposed to do. She responded by telling me to take a couch seat. I showed her my ticket with no results. So I started to complain. My brother was standing near the door and thought I was going to be arrested, so he wrongfully intervened, meaning well. I eventually sat in the club car until we reached Albany, New York, from Toledo, Ohio, which is a far place away. When I got home, I called customer service and got reimbursed. However, my brother had still failed to see my point, thinking only he was right in preventing me for being arrested. I do admit that it was a possibility, but he failed to see my reasoning for making a fuss. I believe that unless you make an issue out of the problem, you can't justify your claim. I will leave it up to you to decide who is correct. I believe both of us.

Another event happened. I had come home, and he wanted me to help him clean the gutters in his house that was close to our parents' home.

It was a hot overcast day. By the end of the day, I was glowing in the dark from looking up all day. When we got back home, both my mother and father busted out laughing; and when I went to the train station to go home and people were looking at me, he had something funny to say

During that same visit, he asked me to look into his storm drain to see if it was plugged. As I was doing it, he started talking to his next-door neighbor with his back turned away from me. Well, wouldn't you know it, I got stuck in my head being partway into the hole. I started calling him for help, to no avail. It took what seemed like an hour for him to turn around and see me. I could have killed him, but we eventually laughed as usual

The last one I want to talk about, even though there are many more, is the time I was staying at his house. He was having a Buddhist meeting. After he introduced me to what I thought was a good friend of his, she and I decided to go to her house for dinner. She said that she hadn't gone to the grocery store yet, and it had started to snow. So my brother gave us a couple of steaks. This should have been a warning. We arrived at her house, me not watching the way she went. After being there for a while with her three children eating the steaks, she said it was time for bed and turned the heat down.

By now I had lost all interest and asked that she drive me home. She responded by saying it was too late now wait until tomorrow and went upstairs to bed, leaving me in the kitchen. So I left thinking I could find my way, having just come with only a spring jacket. Boy, was I wrong. A blizzard had started, and I was without a coat. I must have walked ten miles for a four-mile trip. I figured if I could locate downtown Toledo, I then could find my way home because there was a street that ran from there right past my brother's street. Well, I started out going the wrong way.

When I finally got to the house, my brother just stood there in the doorway, laughing his ass off, saying that I was Santa Claus coming early. Need I tell you it took two days for me to find out. I have never heard the end of that.

These are just a few of the problems being around my brother—never a dull moment—and my mother has always taken anyone's side but mine on anything. My brother says it's because the way I grew up. Too much smoke. I guess in a way, he's right because as you can see, my life has been anything but normal. Does anyone else have a relationship with their brother like me? In case you wanted to ask my relationship with my other baby brother, that's what he is to me, and my two sisters isn't much different. I guess that's why everyone looks for me to come home because there is always something new to laugh about. No one takes any mercy on me, never keeping in mind my difficulty. Finally, I must say he has always been there for me, especially now that I am much older. Without his hounding me and without the help of someone else that I am going to talk about, I wouldn't be able to write this personal story.

CHAPTER 57

Karmai Alexander

This young lady has become very important in my life. We started communicating when I started browsing through Facebook commenting on things of interest. I came upon her very unusual art. I commented on it, saying something like, "I really like it," and that I was happy to see that there is a female black artist trying to bring attention to our heritage. She said thank you, and it went from there to us starting to talk in Messenger. This is where I learned how to use it. From that time on, our friendship grew. We found out what we had things in common, such as having served in Germany driving trucks (of course, at different times); being Aries, which became very important; having a similar hard living background; and much more. As time went on, both being Aries and having been through rough times, we knew that it wasn't easy to trust someone with your personal thoughts. However, we got long past that, growing to a point where we both understand and feel each other's feelings without talking about it. An example is, I have been able to caution her about people and things before they occur. It took some time for us to accept this: her listening and me feeling free to call at any moment of feeling it and her calmly accepting it. I won't reveal the events because they are hers and personal.

On one occasion after we had gotten used to each other, she sent me a picture of some birthday flowers I sent her. I was looking at it and was trying to put it in my folder. But for some reason, it was sent to my brother. I immediately tried to delete it, but it was too late. He was even using his cell phone at that time and called me right away, talking in that curious way, asking me, "What's up with this?" I tried to give him a false story, but he was having none of it. He said, "Now I have proof to show the family that you have a girlfriend, sending her flowers." I pleaded with him to not say anything, but you know how far that got. That is how my family first found out about her. She was there for me when I talked about calling my daughters and pushed me over the hill to actually do it. She has always been there for me, even advancing my need to write this biography. She has been very understanding and, most of all, patient, which has been the most nerve-racking for anyone, especially when you are an artist and used to being isolated to do your best work. I have adjusted to this, but it has been a learning experience. Keep in mind, this relationship has lasted with us not meeting in person yet. So you know that it's real—the most difficult thing I have ever done. It's like God's punishment for my past. But I can deal with it.

She knows how to keep me focused without the need to talk about it. This relationship is very unusual, but I feel that it works for us and isn't that all that matters. We have talked about meeting finally this winter sometime. You know how impatient I am. But seeing I'm not in charge, which is something else I'm not used to, I have to wait for a date of her choice. Now you know that's not the me I even know. But I had finally come to understand that I can't have everything going the way I wish. Man, I can tell you that's a hard lesson to learn. I am giving out her e-mail address and some of her various artwork so if you are interested, you can contact her. She has just opened an entrepreneurship with her sisters, who are also artists—a sign of her growth. Her e-mail is karmaialexander@gmail.com Pictures of her work will be added as appendix 16.

This is just a small part of my important relationship with someone outside of my family. I have God to thank for this, and I truly believe my grandma had a lot to do with this. You have to keep faith.

CHAPTER 58

Information for Seniors

I want to tell you about personal experiences that I have been through. In this way, I hope you will be able to understand how you can handle financial problems. I am focusing my attention on us senior citizens and those who are on a fixed income.

Section 8 housing. When you apply, make sure you have all of your material like doctor material; and if you are a veteran bring all your information because you are entitled to special treatment.

When you have your rectification, you should always bring any out-of-pocket medical experiences like diabetic cream or even dependents all are allowed. Make sure that any increase is correct. Be sure you keep a record on any household problems that weren't attended to properly. If your apartment has been broken into, make a record of all missing items right away so you can have management replace them. Keep a record in some way that you have paid all required bills. Call immediately if there is something that needs to be replaced so you can refer to it if needed. If you have medical problems, be sure to let the management of the property know right away. Be sure you have someone where you live know enough about you so in case of an emergency, you can be helped. Finally, make absolutely sure that you have HUD's number for management problems and your police department number. Recently, the Section 8 tenants have to bring in proof of income every year. If you go to places like CVS, be sure you keep your recipes during the year because they will be needed for proof of expenses. This, in general, are all suggestions.

Bills you pay. Be sure you know what you are paying for and a number that you can call if you have any questions. Many times, you will be charged a few extra dollars considered as tax that really is profit for the company. This most often applies to the elderly because companies rely on the fact that we don't check our monthly bills and do not complain about a few extra dollars. This is highway robbery, and you must have it corrected right away. For example, if your cable bill starts to show an increase that you haven't asked for, call them immediately with your most recent bills in hand. Always ask for a supervisor because the people up front don't want or are trained to confuse you in order to get you off the phone. Never be put on hold because they will not return.

If this happens, immediately call back and ask for "special services." Tell them to listen to what you are calling about before you let them talk. Many times, they say they will look into your problem without knowing what it is. Make sure that if it's a financial one and you have been overcharged, and you have paid already by direct payment, demand your return right away, not an adjustment in your next bill or a two- or three-day repayment. Tell them that they took a minute to take it out, so push the same button to return it because they aren't going to include any interest that they receive from your money. Remember, a penny saved is a penny earned. Keep all information concerning your money on file either by e-mail or paper reports. Written proof is most important if needed for legal reimbursement.

When you go to any office for information, make sure you get a direct understandable answer to your questions, not any goobey-goo. Remember, if they want your money, be sure you know in writing what your rights are because things often change after you start paying. The same people may not be there or act like they don't remember you. Bring a trusted friend with you when you are conducting financial business. It is much safer to have two heads concerning your money because at the place you go to, their first priority is to get you to sign on the dotted line. Always keep that in mind because all that glitters is not gold. When you are enjoying that first cup of coffee in the morning, take time to check your account because if someone has withdrawn money that you didn't authorize, it generally happens at night, and they know that elderly folks don't do this timely in order to get it fixed. Remember, it's your money, and you have none to give away. So checking regularly could save you a lot of stress.

Doctor visits. You must always be able to feel comfortable with whomever you place your life. Ask questions, get references. Don't be shy. Keep in mind that if they are honest, they will be glad to assist you because their reputation is very important to them. Don't be fooled into trying things that are advertised on TV without taking the time to check them out completely and checking it out with your doctor, along with understanding what and how much medication you are taking and for what problems. You have nothing but time, and time means money and your health. Make sure it isn't made at your expense.

Unemployment. If you are entitled to it and are denied, there are three steps for recovery. Read the small print on you material. The last appeal is with the board. You are entitled to be heard. If you win at any level, you should receive your benefits from the time you first applied. Keep in mind, the system isn't designed to help you. It is designed to help make a profit for those that owe you. You can't give up because that's what they expect. When you constantly appear in front of them, being hungry children climbing over everything, you know management will pay you the attention you requested. Let them know if you did the right thing, you wouldn't have to continue here. Offices are never prepared for such actions, and to put you out would create bad publicity for them so they would rather take care of you. Today you have to do whatever is necessary to make sure your rights are protected. You may not always win, but you can always make a point for at least someone else—or maybe get the company to stop doing the wrong thing.

When you receive a call telling you that you have won something, hang up before you start talking because today they can record your voice and, with the right information about you, use it to extract money from your account. When you receive material offering you discounts that you haven't asked for, throw it away Remember, no one just gives you anything without you paying for it at some time. If your cable or any similar company tells you that all you need to pay is a lesser amount than your normal bill, be aware—the following month, there will be interest applied to your next bill. That is what they want. They make more money by you always paying interest. The people handling the phones are only trained to get money for the company There is very seldom a true helpful customer service department. If you noticed, most of the times, that line is busy, trying to get you to think that they are busy. When you want their attention, press any number to be heard and then go directly to the supervisor and let them know that you are a member of AARP. This will often get their attention.

Never—and I mean never—believe that someone you know has submitted your name in order for you to win something. Just hang up. Because common sense will tell you that it's too good to be true. A very important common problem is when you have ordered flowers, and they have messed up. Make sure you have a tracking number, and the phone number of the department that took your money before calling them. By the way, it is also important to check your order a few hours after you purchase your order. In this way, you will immediately know if your order has been sent to the correct person and at the price you have agreed to. When you call to ask questions, demand to speak to a supervisor. More often than not, they are available in the early morning. Tell them your concerns, especially when your delivery is late or it's the wrong order. They are required to immediately correct your order and refund half of your payment. Most want to give you credit for your next order; don't fall for it. If they act like they won't, tell them that you are cancelling the order and demand full reimbursement immediately. Don't fall for any exceptions. Keep in mind, you are not the only one.

If you use Facebook or any other social media, put your problem on, and you will see how quickly they respond because that is one of their major advertising methods. I know this for a fact because I have done it. This is the way to get attention to your issue and get quick action. When you get satisfaction, be sure they send you confirmation and their required $20 deduction on your next purchase. Facebook can be your best friend in cases like this. No company wants bad news about their products.

Finally, when you travel, make sure you ask if there are any benefits for disabled and seniors. Most often, you will find there are but not publicized. To say in general, you must stay alert if you want to survive. The old days of not worrying about not being taken advantage of are gone. Many well-established companies have changed from helping you to helping themselves. All you have to do is look at things today. Do

you think Trump is in office to look after you? Hell no. He
has his agenda to get all he can while he can. It's not his fault;
we knew he was a businessman. So don't believe the hype.
Be smart and stay sharp. Show that you may be old, but not
stupid. Wisdom comes from experience. Who has more of it
than we have? I hope this has been helpful.

In closing, let me say that without the guidance of my
grandmother and the ability to learn from God, I would not
be here today. I am not a churchgoing person, but I believe
in God and His willingness to forgive and allow a person to
redeem themselves. My story should provide a perfect example of His forgiveness. I have seen it all, done most of it all,
and know it all. Now I've shared what I've learned with you
all.

APPENDIX 1

Aunt Eleanor's Achieve Notice

Classification Scheme for Archival Literature

By S. MOOKERJEE

*Librarian, Imperial Record Department
New Delhi, India.*

WRITINGS on Archives and Manuscripts have of late been on the increase. Thanks to the permanent feature of October issues of the AMERICAN ARCHIVIST wherein we are supplied with an excellent and scholarly bibliography of materials of great importance to Archivists and Librarians throughout. The one difficulty which we have been feeling of late is the proper classification of books on Archives and Manuscripts. To all Librarians and Archivists who have libraries attached to their organizations the method of classifying the printed materials on Archives and Records have been presenting some knotty problems. This is more so with Archival Librarians. It is desirable that a uniform scheme based on Dewey's Decimal classification or any other recognized classification scheme having an international acceptance be adopted for the purpose. It is now beyond all contention that Archival knowledge in all its branches comprises not only broad facts of history of the particular region in question but also much technical knowledge about Archives and Archival materials in particular. Knowledge of chemistry and other branches of Science is also deemed necessary for purposes of preservation of Archival materials. The classification regarding phases of Archival Administration formulated by Dr. Posner in the National Archives Staff Information Series No. 12 (May 1942) is of great help to all of us. The list of headings as arranged by Miss Eleanor Walden of National Archives Staff is also a useful suggestion for classifying books and publications on Archives. Hilary Jenkinson's *Manual of Archive Administration* and other books on similar lines have been of great help to us in handling Archival literature in our libraries. In India, to most of us who have been working the Dewey Decimal Classification in the libraries, it has of late been a serious problem to classify the various types of publications that have been pouring in on Archives. This is more so to Archival Libraries which cater to the needs of Archivists and scholars interested in Archives. Subject to final settlement a suggestion will be made in the following lines showing the scheme of classification that we have been working in the Imperial Record Department Library for books on Archival Science. Our Library is mostly a Research Library on Indian History —with a specialisation for the Indo-British period; and Archival literature in particular. We follow the Dewey Decimal Scheme and as such as we have

335

APPENDIX 2

1947 Snowstorm

APPENDIX 3

Rubin James

Reuben James was born in Delaware about 1776. During the Quasi-War with France, Boatswain's Mate James participated in Constellation's victories over the French ships L'Insurgente, 9 February 1799, and La Vengeance.

During the Barbary Wars, James served in Enterprise and accompanied Stephen Decatur into the harbor at Tripoli on 16 February 1804, as Decatur and his men burned the captured American frigate Philadelphia to prevent the Tripolitans from using her in battle. In the ensuing skirmish, an American seaman positioned himself between Decatur and an enemy blade, an act of bravery attributed to Reuben James and Daniel Frazier.

For the rest of the war, James continued to serve Decatur on board Constitution and Congress. During the War of 1812, he served in the United States, under Decatur, and in President. On 15 January 1815, however, President was defeated by the British and James was taken prisoner.

After the war, he resumed service with Decatur, on board Guerriere, and participated in the capture of the 46-gun Algerian flagship Mashonda on 17 June 1815. After peace was made with the Barbary states, James continued his service in the Navy until declining health brought about his retirement in January 1836. He died on 3 December 1838 at the U.S. Naval Hospital in Washington, D.C.

APPENDIX 4

Boston Busing Problem

The desegregation of Boston public schools (1974–1988) was a period in which the Boston Public Schools were under court control to desegregate through a system of busing students. The call for desegregation and the first years of its implementation led to a series of racial protests and riots that brought national attention, particularly from 1974 to 1976. In response to the Massachusetts legislature's enactment of the 1965 Racial Imbalance Act, which ordered the state's public schools to desegregate, W. Arthur Garrity Jr. of the United States District Court for the District of Massachusetts laid out a plan for compulsory busing of students between predominantly white and black areas of the city. The hard control of the desegregation plan lasted for over a decade. It influenced Boston politics and contributed to demographic shifts of Boston's school-age population, leading to a decline of public-school enrollment and white flight to the suburbs. Full control of the desegregation plan was transferred to the Boston School Committee in to; in 2013 the busing system was replaced by one with dramatically reduced busing.

The desegregation of Boston public schools (1974–1988) was a period in which the Boston Public Schools were under court control to desegregate through a system of busing students. The call for desegregation and the first years of its implementation led to a series of racial protests and riots that brought national attention, particularly from 1974 to 1976. In response to the Massachusetts legislature's enactment of the 1965 Racial Imbalance Act, which ordered the state's public schools to desegregate, W. Arthur Garrity Jr. of the United States District Court for the District of Massachusetts laid out a plan for compulsory busing of students between predominantly white and black areas of the city. The hard control of the desegregation plan lasted for over a decade. It influenced Boston politics and contributed to demographic shifts of Boston's school-age population, leading to a decline of public-school enrollment and white flight to the suburbs. Full control of the desegregation plan was transferred to the Boston School Committee in 1988; in 2013 the busing system was replaced by one with dramatically reduced busing.

Appendices Homelessness Metro Program Started 1966.In Boston Massachusetts

My best friend in kindergarten, Eddie Linton, did not live in one of the spacious houses on the hill in the Boston suburb where I grew up in the 1980s and 1990s, Belmont, which is best known for its stellar schools and abundance of Harvard professors. Eddie, who is black, lived instead in a brownstone in the South End of Boston, alongside his two American-born sisters, plus grandparents and aunts and godparents from Barbados, the country where his parents were born.

Every morning, Eddie would get up at 6 a.m. and get on a yellow school bus that took him and dozens of other black kids from Boston to Belmont. He'd spend his school day in Belmont, surrounded by kids who did live in those spacious mansions, and then, at the end of the day, he'd get on the bus and go home. "It was a long day, but my parents wanted me to have exposure to a better education system," he told me recently. While he was gazing out the bus window, watching the scenery change from suburban to urban, wealthy to middle- and low-income, thousands of other black kids across Boston were sitting on similar buses that took them to and from schools in other predominantly white suburbs, such as Newton, Sharon, and Wellesley, areas that white families had embraced to escape the city in the 1960s and 1970s.

Eddie was a participant in the Metropolitan Council for Educational Opportunity program, one of the longest-running voluntary school-integration programs in the country. Started in 1966, METCO has bused thousands of students in Massachusetts—at least 200 in the first decade to 3,000 since the 1970s—from predominantly black and Latino neighborhoods in the city of Boston and later Springfield to white, wealthy neighborhoods in the suburbs.

The original idea behind the program was to help black kids access better educational opportunities than those available in Boston, and to give white students in suburbia the opportunity to "share a learning experience with students with differing social, economic, and racial backgrounds," as program backers put it at the time. Its founders assumed that it wouldn't be necessary for long—soon, they hoped, housing segregation would dissipate and schools would be places where black and white students were educated alongside one another, without any busing necessary. Homelessness is the condition of people lacking "a fixed, regular, and adequate nighttime residence" as defined by The McKinney–Vento Homeless Assistance Act. According to the US Department of Housing and Urban Development's Annual Homeless Assessment Report, as of 2018 there were around 553,000 homeless people in the United States on a given night, or 0.17% of the population.

Homelessness emerged as a national issue in the 1870s. Many homeless people lived in emerging urban cities, such as New York City. Into the 20th century, the Great Depression of the 1930s caused a devastating epidemic of poverty, hunger, and homelessness. There were two million homeless people migrating across the United States. In the 1960s, the deinstitutionalization of patients from state psychiatric hospitals, according to the physician's medical libraries on use of pharmaceuticals, was a precipitating factor which seeded the population of people that are homeless.

The number of homeless people grew in the 1980s, as housing and social service cuts increased. After many years of advocacy and numerous revisions, President Ronald Reagan signed into law the McKinney–Vento Homeless Assistance Act in 1987; this remains the only piece of federal legislation that allocates funding to the direct service of homeless people. Over the past decades, the availability and quality of data on homelessness has improved considerably. About 1.56 million people, or about 0.5% of the U.S. population, used an emergency shelter or transitional housing program between October 1, 2008 and September 30, 2009. Homelessness in the United States increased after the Great Recession.

In the year 2009, one out of 50 children or 1.5 million children in the United States of America will be homeless each year. There were an estimated 37,878 homeless veterans estimated in the United States during January 2017, or 8.6 percent of all homeless adults. Just over 90 percent of homeless U.S. veterans are male Texas, California and Florida have the highest numbers of unaccompanied homeless youth under the age of 18, comprising 58% of the total homeless under 18 youth population. Homelessness affects men more than women. In the United States, about 60% of all homeless people are men. However, about 71% of all homeless individuals are male.

Because of turnover in the population of people that are homeless, the total number of people who experience homelessness for at least a few nights during the course of a year is thought to be considerably higher than point-in-time counts. A 2000 study estimated the number of such people to be between 2.3 million and 3.5 million. According to Amnesty International USA, vacant houses outnumber homeless people by five times December 2017 investigation by Philip Alston, the U.N. Special Rapporteur on extreme poverty and human rights, found that homeless persons have effectively been criminalized throughout many cities in the United States.

Causes of homelessness in the United States include lack of affordable housing, divorce, lawful eviction, negative cash flow, post traumatic stress disorder, foreclosure, fire, natural disasters (hurricane, earthquake, or flood), mental illness, physical disability, having no family or supportive relatives, substance abuse, lack of needed services, elimination of pensions and unemployment entitlements, no or inadequate income sources (such as Social Security, stock dividends, or annuity), poverty (no net worth), gambling, unemployment, and low-paying jobs. Homelessness in the United States affects many segments of the population, including families, children, domestic violence victims, ex-convicts, veterans, and the aged. Efforts to assist the homeless include federal legislation, non-profit efforts, increased access to healthcare services, supportive housing, and affordable housing.

APPENDIX 5

Veteran Benefits

The Bonus Army were the 43,000 marchers—17,000 U.S. World War I veterans, their families, and affiliated groups—who gathered in Washington, D.C. in mid-1932 to demand cash-payment redemption of their service certificates. Organizers called the demonstrators "Bonus Expeditionary Force," to echo the name of World War I's American Expeditionary Forces, while the media referred to them as the "Bonus Army" or "Bonus Marchers." The contingent was led by Walter W. Waters, a former sergeant.

Many of the war veterans had been out of work since the beginning of the Great Depression. The World War Adjusted Compensation Act of 1924 had awarded them bonuses in the form of certificates they could not redeem until 1945. Each certificate, issued to a qualified veteran soldier, bore a face value equal to the soldier's promised payment compound interest. The principal demand of the Bonus Army was the immediate cash payment of their certificates.

On July 28, U.S. Attorney General William D. Mitchell ordered the veterans removed from all government property. Washington police met with resistance, shots were fired, and two veterans were wounded and later died. President Herbert

Hoover then ordered the Army to clear the marchers' camp-site. Army Chief of Staff General Douglas MacArthur commanded the infantry and cavalry supported by six tanks. The Bonus Army marchers with their wives and children were driven out, and their shelters and belongings burned.

A second, smaller Bonus March in 1933 at the start of the Roosevelt administration was defused in May with an offer of jobs with the Civilian Conservation Corps at Fort Hunt, Virginia, which most of the group accepted. Those who chose not to work for the CCC by the May 22 deadline were given transportation home.[1] In 1936, Congress overrode President Roosevelt's veto and paid the veterans their bonus nine years early.

Bonus Army Conflict

Bonus Army marchers (left) confront the police.

APPENDIX 6

Munich Airplane Crash

On 17 December 1960, the Samaritan was due to fly from Munich-Riem airport in Germany to RAF Northolt in the United Kingdom with 13 passengers and 7 crew. Shortly after takeoff, the aircraft lost power to one of its two Pratt & Whitney R-2800radial engines. Unable to maintain altitude, it hit the 318-foot steeple of St. Paul's Church next to the Oktoberfest site (then vacant) in the downtown Ludwigsvorstadt borough. Subsequently, at 2:10 PM, it crashed into a crowded two-section Munich tramway car in Martin-Greif-Straße, close to Bayerstraße.

All 13 passengers and 7 crew members on the plane died. 32 people on the ground were killed and 20 were injured. A section of the wing crashed through the roof of a building at Hermann-Lingg-Straße, a block away from the main accident site, without injuring anybody there. Time magazine later reported that all 13 passengers on the Convair were holiday-bound

APPENDIX 7

LaGuardia Airport Bombing

On December 29, 1975, a bomb was detonated near the TWA baggage reclaim terminal at LaGuardia Airport, New York City. The blast killed 11 people and seriously injured 74. The perpetrators were never identified, although investigators and historians believe that Croatian nationalists were the most likely. The attack occurred during a four-year period of heightened terrorist attacks within the United States. 1975 was especially volatile, with bombings in New York City and Washington D.C. early that year and two assassination attempts on US President Gerald Ford.

The LaGuardia Airport bomb, at the time, was the single most deadly attack by a non-state actor to occur on American soil since the Bath School bombings, which killed 44 people in 1927. It was the deadliest attack in New York City since the Wall Street bombing of 1920, which killed 38, until the September 11 attacks in 2001.

APPENDIX 8

Union Staff and My Affiliation with the National Union of Hospital and Health-Care Employees

This appendix includes pictures of me seated at the AFL/CIO executive board meeting in 1975. Also pictures of me as district president and a statement from Boston Globe.

Number of workers organized in 1973: 173 Number of workers organized in 1974: 304

OVER-VI Prior to November, there was an attitude that "As St. Elizabeth's went, so went 1199MASS," I think that attitude has been corrected and we have begun serious organization in Massachusetts outside of Boston and should see some results soon. We have good possibilities of winning the St. Elizabeth's campaign but there will still be difficulties in organizing in Boston, especially now because of the divisions caused by the bussing issue. Staff turnover has been a problem but in the last 3 months we have stabilized the staff and should see some progress in Massachusetts.

Status of Campaigns as of November: At the Executive Board meeting in November 1914, a great deal of weight was placed on the success of the campaign at St Elizabeth's Hospital in Boston. We have maintained a strong organization at this major institution and should be able to win the NLRB Service and Maintenance unit.

Campaigns Started Since November: We have begun very good campaigns in both Fairlawn Hospital in Wooster, Mass. and Franklin County Hospital near Springfield. We have also begun a campaign at St. John's Hospital in Lowell but it is too early to know what the possibilities are there.

We have started organization in several nursing homes.

Since November we have won elections at one nursing home and two abortion clinics and have just finished hearings at another nursing home where we expect to win an election soon.

We have had serious problems with the organization of day care centers, especially in terms of negotiating contracts and with the financial stability of the centers. Elliot Small has prepared a detailed report for the National Executive Board on the organization of day care centers.

Staff: The staff consists of Elliot Small, the Area Director, and 7 organizers. Ken Walden is responsible for the administration of the large institutions in Boston. Tom White is responsible for the administration of nursing homes in Boston. Mary Clemmens is responsible for all aspects of the day care center organization. Alneta Bond is organizing in Boston. Ken Allen is currently helping out at Franklin County Hospital in Springfield and will be working in organization on the North Shore. Jim Bollen is responsible for administration and organization on the North Shore. Lillian Viens is responsible for organization on the South Shore Cape Cod area. She has several nursing homes under serious organization. Vincent Griesi resigned in January and we have found a replacement for him in Springfield starting in mid May. Elliot Small and Ken Allen are responsible for the St. Elizabeth's Hospital campaign.

Unions expect improvement in recruiting

By Nils Bruzelius
Globe Staff

Spokesmen for two unions that have tried and failed to organize Beth Israel Hospital employees in the past were pleased about yesterday's Supreme Court decision affirming the right of hospital employees to solicit and distribute union literature in the hospital cafeteria and coffee shop.

The decision will make a psychological difference in future union organizing efforts, said Stephen Masur, a union representative of the Hospital Workers Local 880 of the Service Employees International Union, AFL-CIO.

It was Local 880 that in 1973 filed unfair labor practices charges against the Brookline avenue hospital after it warned an employee that she would be fired if she distributed union literature in the contested areas.

In arguments before the court, the hospital had argued that since patients and visitors also used the disputed areas, union solicitation there could interfere with the hospital's wish to provide a proper recuperative environment.

Spokesman Anthony Lloyd said the hospital would not comment until it had received and read the lengthy court decision.

He noted that the hospital's rule against union solicitation in the cafeteria and coffee shop has remained in force throughout the appeals process, which went against the hospital at each stage. The hospital has allowed solicitation only in locker rooms and rest rooms.

"Just on the face of it," Masur said, "it doesn't sound as if it's as broad a decision as we would have liked."

He said, however, that the union will now feel free to encourage union solicitation by employees in the cafeteria and coffee shop.

"We've been obeying the (hospital's) rule, but we now assume the rule is invalid," he said.

Being able to use the cafeteria and coffee shop, he said, "would make a psychological difference, because union activity would be seen by people as a little more legitimate."

Kenneth E. Walden, secretary-treasurer of District 1199 of the National Union of Hospital and Health Care Employees union, AFL-CIO, called the Supreme Court decision a minor victory.

"I'm pleased," he said, "but I just feel that it's a decision without any teeth in it."

In February, Local 880 lost decisively in a supervised election to determine whether the hospital's service and maintenance employees wanted the union to represent them.

Hospital employees similarly rejected a union representation in a 1974 election in which both unions participated.

Pictures of me: 1) Veterans' shelters, (2)
UMass, Boston, (3) Veterans' training

APPENDIX 9

My Veterans ID, My UMASS ID, and My Veterans College ID My UMASS College Degree, My Veterans Upward Bounds Certificate

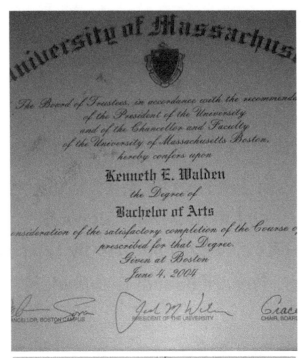

APPENDIX 10

Dr. Robert Walden History

Obituary for Robert Walden M.D.

Col. Robert E. Walden, M.D. (ret.) passed from this earth peacefully in his sleep at 5:45 a.m. on Friday, October 12, 2012 at the age of 92. His health failed in May of this year and he succumbed to illness after a brave and lengthy struggle.

Dr. Walden was a founding faculty member of the Medical College of Ohio. He moved to Toledo in 1968 to help create the curriculum for the charter class of medical students, and moved his family here a year later. As Director of Psychiatric Inpatient Services at MCO, Dr. Walden developed the college's first clinical psychiatric unit. He served on the college's Affirmative Action and Admissions committees. And, he was an Adjunct Assoc. Professor of Psychiatry at the Univ. of Toledo. He retired from MCO as an Emeritus Professor of Psychiatry in 1990.

He was a 1st Lieutenant in the USAR Medical Corps during WWII, and Captain in the USAFR Medical Corps following the Korean War. He continued his service to this country and our veterans while working for the Veterans Hospital Service in civilian life. And, he returned to military service as a Lt. Colonel in the USAR Medical Corps in the 1970's. He also traveled to the Soviet Union as a civilian medical ambassador during the Cold War. He retired from the military and government service in 1984 at the rank of Colonel.

Dr. Walden was a trailblazer in the fields of mental health and community medicine. He helped to humanize and reduce the number of chronically-ill psychiatric inpatients by introducing innovative, humane therapy to their treatment. He was also a strong advocate of reintroducing patients to society.

He was the founder and project director of the Cordelia Martin Health Center on W. Bancroft St. in 1969. And, he pioneered health screening programs under the Model Cities program in the 1970s.

A graduate of Lincoln University (PA) and Meharry Medical College (TN), Dr. Walden was a Life Member of Kappa Alpha Psi Fraternity, and a founding member of Sigma Pi Phi Fraternity—Alpha Phi Boulé. He is Past-President of the Ohio Psychiatric Assoc., and was an international lecturer; presenting papers in the Philippines, Japan, and the U.S. He's listed in Who's Who in America.

Dr. Walden was an active Congregationalist, and gave generously to the church throughout his life. He served as a deacon and trustee of First Congregational Church of Toledo. And, he was active in the National Assoc. of Congregational Christian Churches leadership. He is survived by his wife of 59 years, Ethel L. Walden; his children, Kenneth Walden, Roberta Miller, Robert Walden Jr., Dr. Mark Walden and Mary Mitchell, Esq.; 9 grandchildren, 4 great grandchildren, and many nieces and nephews.

Visitation is Friday, Oct. 19, 5-8 p.m. at Dale-Riggs Funeral Home, 572 Nebraska Ave. at City Park. The funeral is Saturday, Oct. 20, 2 p.m. at Mayfair-Plymouth Congregational Church, 5253 Bennett Rd. near Alexis. And, charitable gifts can be made in his name to the Lincoln University or Meharry Medical College Foundations.

To send flowers or a memorial gift to the family of Robert Walden M.D. please visit our Sympathy Store.
© 2019 Dale-R

APPENDIX 11

Pension Fund Problems

This appendix contains all correspondence concerning my pension fund.

National Union of Hospital and Health-Care Employee Union Pension Fund Problems

> Harvey Law Offices
> November 22, 2016
> George Gresham, President of 1199 SEIU, United HealthCare Workers East 310 West 43rd Street New York, NY 10036-6407
>
> Henry Nicholas, President 1199C National Union of Hospital and Health Care Employees 1319 Locust St Philadelphia, PA 19107
>
> Re:
> Kenneth Walden
> Gentlemen:

I represent Kenneth Walden of Cambridge, Massachusetts. As you are aware from previous correspondence from Mr. Walden, he played a key role in establishing the Union in Boston back in 1964. He has been greatly dismayed at the portrayal of the hi history as outlined in the Union's website, which completely omits the contributions that he and the Boston local made to the Union's success. Moreover, Mr. Walden, a pioneer on this front, is personally distressed to see Tyrek D. Lee identified as the first African-American man to lead a major statewide union in Massachusetts, an honor Mr. Walden has held for decades. He certainly holds no animosity toward Mr. Lee and in fact is delighted to see people of color succeed. However, Mr. Lee wasn't even born when Mr. Walden was breaking barriers and growing the Union and he art in the Union's history acknowledged. Additionally, he would like to be compensated for his role in the establishment of the Union.

History

Back in 1964 Mr. Walden and three of his co-the Jewish Memorial Hospital in Roxbury, Massachusetts, began the Boston Healthcare Workers Union. These employees worked under harsh conditions, including working more than 44 hours per week with no over-time. At a time of great civil unrest as well as violent racial tensions in the US, they worked together across racial divides and organized scores of nursing homes throughout the state. Once hospital employees were able to unionize, Mr. Walden and his associates brought hundreds more employees into the Union. By 1968 the Boston Healthcare Workers joined National Union of Hospital and Healthcare Employees, AFL/CIO, in New York. At that time, Elliott Small of New York joined the local as President and Mr. Walden was Secretary Treasurer. In that role, he was charged with negotiating all contracts in the state. Mr. Walden also served on the National

Helping families prepare for the inevitable, protect against the unexpected and preserve their savings.

George Gresham Henry Nicholas November 22, 2016 Page 2

Union's Executive B howard as Vice President throughout the 1970's. When Boston became a District in 1977, he was elected the first President—the first African American president as well—an office in which he served until he was forced out in 1980. During his time on the National American Arbitration Advisory Board from 1972–1980, Mr. Walden convinced the Board to allow a Union officer to handle arbitrations. He knew that no one understood the contract better than the Union and its members. He also conducted trainings for Shop Stewards to conduct Grievances in order to provide the groundwork for successful litigation if needed. These innovations, while considered the norm now, broke new ground at the time. While serving the National Executive Board as District Vice President, Mr. Walden served with Leon Davis, Doris Turner and Henry Nicholas in 1979 to establish the new Union By-Laws in Pawling, New York. Although he later became an important part of the Executive Board, at that time Bob Molenkamp had no official role with the Union but only acted as assistant to Leon Davis.

Mr. Walden negotiated the Union's first contract with Boston University Hospital (now Boston Medical Center) and its subsequent contracts through 1980—resulting in a more than $10.00 per hour increase in member wages. In fact, Mr. Walden was so respected by hospital management teams that he was asked on several occasions to speak with management organizations about contract terms and the union's perspective on negotiations. According to Mr. Walden, Henry Nicholas met him at Logan Airport in Boston in 1980, shortly after he had closed negotiations on the BU Hospital contract. He offered Mr.

Walden $54,000 to reopen the contract negotiations to include the Benefit Fund. Mr. Walden refused. Shortly thereafter, his relationship with the national union became strained. The Union officials cut his funding in 1980 and refused to support him in handling contract negotiations and other Union business, including arbitrations. forced him out without that month's pay. The Union then changed its affiliation to SEIU and Henry Nicholas became the National Union President.

Mr. Walden was constructively terminated from the Union. He lost his job, his family and all his belongings as a result. He was unemployed for the first time in his working life. He spent time at the Pine Street Inn, a homeless shelter in Boston, after his termination. It took him many years of hard work to regain his position in the community that he lost as a result of the Union's wrongful treatment of him.

Requests

Mr. Walden would like an apology from the Union for its mistreatment of him. He requests that the Union's website and history be revised to include reference to him and his
embers. He would also like to apply for the pension benefits that he is entitled to based on his service as a union officer

from 1968 through 1980. Please forward the appropriate pension documentation to me at my Salem office address.

I look forward to hearing from you so that Mr. Walden can finally have these issues resolved.

Thank you.

THiDenise Leydon Harv City and Washington D.C. Wednesday, December 13, 2017

Head of SEIU 1199 in Massachusetts Is Suspended Over "Inappropriate Conduct"

Tyrék D. Lee Sr., the top official at SEIU Local 1199 in Massachusetts, has been suspended over "inappropriate conduct," according to the Boston Globe.

The newspaper cites unnamed "people familiar with the situation" who say Lee was suspended over "accusations of sexual harassment." The local represents approximately 56,000 healthcare workers in Massachusetts. (Priyanka Dayal McCluskey, "Head of health care union suspended over allegations of inappropriate behavior," Boston Globe, December 12, 2017)

Lee holds the title of "Executive Vice President" at SEIU Local 1199, where last year he earned $128,902, according to records from the US Department of Labor.

Union officials did not detail allegations against Lee, said the Globe. Here's what SEIU Local 1199 said in a statement to the press:

"1199SEIU strongly condemns all forms of inappropriate conduct and will not tolerate such behavior by any employee of our union. Upon being made aware of these allegations 1199SEIU has taken the action of suspending Executive Vice President Tyrék Lee while a formal investigation is conducted."

The Globe notes that "Lee took the top job at 1199SEIU in January 2016, when he was 38."

In November, SEIU appointed SEIU Executive Vice President Leslie Frane to lead an internal investigation due to the multiple SEIU officials accused of sexual harassment and/or misconduct, including the suspension and subsequent resignation of SEIU Executive Vice President Scott Courtney.

SEIU also announced the formation of an external advisory group including Cecilia Muñoz, former White House Domestic Policy Council director; Fatima Goss Graves, president and CEO of the National Women's Law Center; and employment attorney Debra Katz, founding partner of law firm Katz Marshall & Banks.

It's possible Lee's suspension was connected to the investigation by Frane's team.

If so, this may create some nervousness over at SEIU-UHW, where staffers and members at SEIU-UHW describe a culture of bullying and sexual misconduct, including alleged affairs by the union's president, Dave Regan. In mid-November, Marcus Hatcher, one of the union's top officials, was fired for allegedly having simultaneous affairs with three female members of the union's Executive Board.

Will Frane's crew soon be knocking on Regan's door?

PENSION ACTION CENTER, GERONTOLOGY INSTITUTE
MCCORMACK GRADUATE SCHOOL OF POLICY AND GLOBAL STUDIES
UNIVERSITY OF MASSACHUSETTS BOSTON

100 Morrissey Boulevard
Boston, MA 02125-3393
P: 617.287.7307
F: 617.287.7080
www.umb.edu/pensionaction

December 19, 2018

Kenneth Walden
365 Rindge Avenue, Apt. 10B
Cambridge, MA 02140

Dear Mr. Walden:

We have not yet received a response from the SEIU 1199 Pension Fund office in Philadelphia. Legally, they should have responded long ago.

Today I sent the enclosed letter reminding them of their responsibility. I also requested a copy of the pension plan that would apply to you.

I'll let you know when I receive a response.

In the meanwhile, have a lovely Christmas. May 2019 bring joy, health, and good news!

Sincerely,

Susan Hart
Pension Counselor

Enclosure: letter to SEIU Philly

Important Notice: Please keep these letters for your records.

In accordance with the Pension Action Center's file retention policy, files will be maintained for a minimum of seven years from the conclusion of representation. After seven years, files will be destroyed in a manner that ensures client confidentiality.

PENSION ACTION CENTER, GERONTOLOGY INSTITUTE
MCCORMACK GRADUATE SCHOOL OF POLICY AND GLOBAL STUDIES
UNIVERSITY OF MASSACHUSETTS BOSTON

100 Morr
Boston, M
P: 617.28
F: 617.287
www.umb.ec

September 28, 2018

BY CERTIFIED MAIL; RETURN RECEIPT REQUESTED

Gwen Watson, Pension Administrator
1199 SIEU United Health Care Workers
1319 Locust Street
Philadelphia, PA 19107

Re: Kenneth Walden   Soc. Sec. No: xxx-xx-6907
      364 Rindge Avenue, Apt. 10B   D/O/B: 04/03/1942
      Cambridge, MA 02140

Dear Sir or Madam,

Please be advised that Kenneth Walden has requested the assistance of the New England Pen
Action Project with respect to the issue of his entitlement to a pension pursuant to his employment by
National Union of Hospital and Healthcare Workers District 1199.

Mr. Walden and three co-workers began the Boston Healthcare Workers Union in 1964. Th
BHW became part of the National Union of Hospital and Healthcare Employees 1199, AFL-CIO, loc
in New York in 1968---the same year that the Benefits, Pension, and Training Fund was established.
Walden served as Secretary Treasurer and International Vice President of the NUHHCE, on the Nati
Arbitration Advisory Board, and as the first President of the Boston District. He was instrumental in
drawing up of the National Union By-laws in 1979.

Please provide me with a statement of Mr. Walden's pension vesting and service credit accru
and, if applicable, a statement of his accrued benefit pursuant to the 1199 SIEU United Health Care
Workers Pension Plan along with a copy of any worksheet and any other documents used to arrive at
benefit amount.

You will find, enclosed with this letter, a signed Release from Mr. Walden authorizing
disclosure of this information to the New England Pension Assistance. Please direct your response to
at: Susan Hart, New England Pension Assistance Project, Gerontology Institute, UMass Boston,
Morrissey Blvd., Boston, MA 02125.

This request is made pursuant to §§ 104(b) (4) and 105(a) of the Employee Retirement In
Security Act of 1974 ("ERISA"). This letter is not a claim for benefits and should not be construed as
Pursuant to ERISA § 502(c), we are entitled to a written response within 30 days of this letter. Than
or your timely response to this request.

Sincerely,

Susan Hart
Pension Counselor

losures: release, IES, letters
: Walden

I AM WHO I AM

CRTR2709-CR

COMMONWEALTH OF MASSACHUSETTS
SUFFOLK COUNTY CIVIL
Docket Report

9384CV04108 Walden v Olmstead

CASE TYPE:	Administrative Civil Actions	FILE DATE:	07/09/1993
ACTION CODE:	E17	CASE TRACK:	A - Average
DESCRIPTION:	Civil Rights Act, G.L. c. 12 § 11H		
CASE DISPOSITION DATE	02/12/1996	CASE STATUS:	Open
CASE DISPOSITION:	Judgment after Non- Jury Trial	STATUS DATE:	02/12/1996
CASE JUDGE:		CASE SESSION:	Civil G

PARTIES

Plaintiff
Walden, Kenneth	Private Counsel
Denise Leydon Harvey
Harvey Law Offices
Harvey Law Offices
27 Congress St
Suite 305 18
Salem, MA 01970
Work Phone (978) 745-5610
Added Date: 07/09/1993	55309
Defendant
Olmstead, Peter	Private Counsel
Linda Shane Olmstead
Law Office Of Linda Olmstead
Law Office Of Linda Olmstead
705 Centre St
Jamaica Plain, MA 02130
Work Phone (617) 524-2821
Added Date: 08/26/1993	37884

Docket Report

INFORMATIONAL DOCKET ENTRIES

	Ref	Description	Judg
1993	1	Complaint & Jury demand on Complaint filed	
1993		Origin 1, Type E17, Track A.	
1993	2	Civil action cover sheet re: complaint	
1993	3	Motion for attachment of real estate by Kenneth Walden & Denied. (Garsh, J.)	
1993	4	SERVC RETRND: Peter Olmstead (Defendant) by leaving at last and usual place of abode Aug. 5,1993.	
1993	5	ANSWER: Peter Olmstead (Defendant)	
1993		Case status changed to "Needs review for answers" at service deadline review	
1993		Case status changed to "Needs discovery" at answer deadline review	
1994	6	Motion for summary judgment by plff (with partial opposition)	
1994		Motion (P#6) Allowed was to count II Denied as to Count I and III (Houston, J) notice sent 7-6-94	
1995	7	JUDGMENT OF DISMISSAL: The complaint of plff is Dismissed for lack of prosecution without costes entered on docket pursuant to Mass. R. Civ.P. 58(a) And notice sent to parties pursuant to Mass.R. Civ.P. 77(d)	
1995	8	FINDINGS ON ASSESSMENT OF DAMAGES (Doerfer,J) Notice sent 07/19/95	
2/1996	9	JUDGMENT ON FINDINGS BY THE COURT AFTER SUMMARY JUDGMENT HAVING BEEN ALLOWED AND ASSESSMENT OF DAMAGES HAVING BEEN RENDERED: 1. Damages are assessed in the amount of $50,000.00 and approve (Doerfer, J.) (filed 7/19/95)., entered on Docket pursuant to Mass.R.Civ.P.58(a) and notice sent to parties pursuant to Mass.R.Civ.P.77(d)	
2/1996		Judgment for Plff - Damages $66,681.00 - Costs $68.00 - Execution issued April 12, 1996.	

APPENDIX 12

Homelessness

Homelessness is the condition of people lacking "a fixed, regular, and adequate nighttime residence" as defined by The McKinney–Vento Homeless Assistance Act. According to the US Department of Housing and Urban Development's Annual Homeless Assessment Report, as of 2017 there were around 554,000 homeless people in the United States on a given night, or 0.17% of the population.

Homelessness emerged as a national issue in the 1870s. Many homeless people lived in emerging urban cities, such as New York City. Into the 20th century, the Great Depression of the 1930s caused a devastating epidemic of poverty, hunger, and homelessness. There were two million homeless people migrating across the United States. In the 1960s, the deinstitutionalization of patients from state psychiatric hospitals, according to the physician's medical libraries on use of pharmaceuticals, was a precipitating factor which seeded the population of people that are homeless.

The number of homeless people grew in the 1980s, as housing and social service cuts increased. After many years of advocacy and numerous revisions, President Ronald Reagan signed into law the McKinney–Vento Homeless Assistance

Act in 1987; this remains the only piece of federal legislation that allocates funding to the direct service of homeless people. Over the past decades, the availability and quality of data on homelessness has improved considerably. About 1.56 million people, or about 0.5% of the U.S. population, used an emergency shelter or transitional housing program between October 1, 2008 and September 30, 2009. Homelessness in the United States increased after the Great Recession.

In the year 2009, one out of 50 children or 1.5 million children in the United States of America will be homeless each year. There were an estimated 57,849 homeless veterans estimated in the United States during January 2013, or 12 percent of all homeless adults. Just under 8 percent of homeless U.S. veterans are female. Texas, California and Florida have the highest numbers of unaccompanied homeless youth under the age of 18, comprising 58% of the total homeless under 18 youth population. Homelessness affects men more than women. In the United States, about 60% of all homeless people are men. However, about 71% of all homeless individuals are male.

Because of turnover in the population of people that are homeless, the total number of people who experience homelessness for at least a few nights during the course of a year is thought to be considerably higher than point-in-time counts. A 2000 study estimated the number of such people to be between 2.3 million and 3.5 million. According to Amnesty International USA, vacant houses outnumber homeless people by five times December 2017 investigation by Philip Alston, the U.N. Special Rapporteur on extreme poverty and human rights, found that homeless persons have effectively been criminalized throughout many cities in the United States.

Causes of homelessness in the United States include lack of affordable housing, divorce, lawful eviction, negative

I AM WHO I AM

cash flow, post traumatic stress disorder, foreclosure, fire, natural disasters (hurricane, earthquake, or flood), mental illness, physical disability, having no family or supportive relatives, substance abuse, lack of needed services, elimination of pensions and unemployment entitlements, no or inadequate income sources (such as Social Security, stock dividends, or annuity), poverty (no net worth), gambling, unemployment, and low-paying jobs. Homelessness in the United States affects many segments of the population, including families, children, domestic violence victims, ex-convicts, veterans, and the aged. Efforts to assist the homeless include federal legislation, non-profit efforts, increased access to healthcare services, supportive housing, and affordable housing.

APPENDIX 13

Pictures of my Artist Friend's Work

These are some of my friend's various artwork. The first one is the picture that was sent to my brother by mistake.

I AM WHO I AM

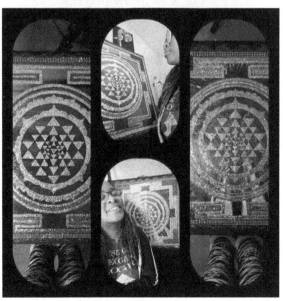

Family Pictures

Here is a picture of our mother with us at her nine-ty-ninth-year birthday party. From left to right: my brother Robert, my sister Roberta, me, my sister Mary, and my brother Mark.

My Leah and my grandchildren Lauren and Rayen.

This is the only picture that I have with me and
my two beautiful daughters together.

ABOUT THE AUTHOR

Kenneth Walden was born in Roxbury, Massachusetts, on April 3, 1942, and raised by his grandmother (born 1875). He was guided by her teaching about the real slavery problems and taught how to survive and maintain self-respect. He started working at the age of twelve and had his own apartment at fifteen. He went into the Air Force twice in 1958 and was sent home in 1957. He went to Germany until 1961, when he was given an honorable discharge. He was involved in civil rights movement throughout the '60s and '70s and was deeply involved in the Boston busing demonstrations. He was a board member of the METCO Program. He took part in the fight to stop eminent domain and was cofounder of the National Union of Hospital and Healthcare employees Union district 1199 Massachusetts in 1966/ 1980. He was the first black president of this organization. He worked at Suffolk Superior Court Office. He experienced being homeless. He is a college graduate with a BA degree. He was the past president of Cambridge Economic Opportunity Commission and now stays home, disabled.

CPSIA information can be obtained
at www.ICGtesting.com
Printed in the USA
BVHW071730060421
604336BV00003B/267

9 781098 031763